NEW FEATURES ON MAGNESIUM ALLOYS

Edited by **Waldemar Alfredo Monteiro**

New Features on Magnesium Alloys
http://dx.doi.org/10.5772/2810
Edited by Waldemar Alfredo Monteiro

Contributors

Zhifeng Wang, Weimin Zhao, Tomasz Tański, Anna Dobrzańska-Danikiewicz, Tomasz Tański, Szymon Malara, Justyna Domagała-Dubiel, Nina Angrisani, Jan-Marten Seitz, Andrea Meyer-Lindenberg, Janin Reifenrath, Masafumi Noda, Yoshihito Kawamura, Tsuyoshi Mayama, Kunio Funami, Parviz Asadi, Kamel Kazemi-Choobi, Amin Elhami, W.A. Monteiro, S.J. Buso, L.V. da Silva

Published by InTech
Janeza Trdine 9, 51000 Rijeka, Croatia

Publishing Process Manager Iva Simcic
Technical Editor InTech Prepress, Novi Sad
Cover InTech Design team

Additional hard copies can be obtained from orders@intechopen.com

New Features on Magnesium Alloys, Edited by Waldemar Alfredo Monteiro
p. cm.
ISBN 978-953-51-0668-5

For EU product safety concerns:
IN TECH d.o.o., Prolaz Marije Krucifikse Kozulić 3, 51000 Rijeka, Croatia,
info@intechopen.com or visit our website at intechopen.com.

Meet the editor

Professor Dr Waldemar Alfredo Monteiro, physicist. He received his MSc (Solid State Physics), DSc (Nuclear Technology) at University of São Paulo, São Paulo, Brazil. Materials Engineering Specialty (Brazilian Physical Researches Center, Rio de Janeiro, Brazil); Physical Metallurgy Specialty (Max Planck Institut für Metallkunde, MPI, Stuttgart, Germany). Dr Monteiro's expertise areas are Physical Metallurgy, Powder Metallurgy and Materials Characterization (Optical and Electron Microscopy; Microanalysis Techniques). He is productivity fellowship researcher (CNPq – Brazil); senior researcher in metallic materials at Nuclear and Energetic Researches Institute, São Paulo, Brazil; Professor (graduate course) on Materials Characterization (Electron Microscopy Techniques and Microanalysis); researcher and lecturer (undergraduate and graduate courses) at Presbyterian Mackenzie University (São Paulo, Brazil). He published one book, two book chapters and 150 articles (scientific journal or congress proceedings). The academic advisories include 30 scientific initiation (under graduate students), 22 MSc and 14 DSc.

Contents

Preface

Magnesium alloys have always been attractive to designers due to their low density, only two thirds that of aluminium. This has been a major factor in the widespread use of magnesium alloy castings and wrought products. The present book gives us an overview in some special areas of magnesium alloys concerning technological applications and eco-friendly requirements.

Each chapter brings us a new facet relating to the application of magnesium alloy including, as an example, researches with rare earth metals as alloying components in magnesium implants for orthopaedic applications.

Quasicrystals (QCs) are a well-defined ordered phase of solid matter with long-range quasiperiodic translational order and an orientational order, but no three dimensional translational periodicity. QC master alloys can be used to strengthen magnesium alloys (surface coating application for frying pans, surgical blades and hydrogen storage materials).

A further requirement in recent years has been for superior corrosion performance and intense improvements have been demonstrated for new magnesium alloys. Mechanical properties enhancements and corrosion resistance have led to greater interest in magnesium alloys also for aerospace and specialty applications.

The galvanic corrosion problem can only be solved by proper coating systems. Magnesium and its alloys form a very thin surface film which indicates that the underlying metal cannot be completely covered. As a result, magnesium and its alloys are highly susceptible to corrosion. In the past, the corrosion behavior of magnesium alloys is the overriding factor to prevent their applications.

Some properties are of interest only for the surface of the material, investigations were carried out concerning surface treatment of the magnesium alloys by applying of the physical vapor deposition processes.

The best development and application prospects tentative from an analysis of mechanical properties of the casting magnesium alloys subjected to laser treatment are exhibited by those materials into which the particles of titanium and vanadium carbides were cladded.

An effective way to improve the ductility of the special magnesium alloys while maintaining high strength involves ensuring recovery of the secondary long-period stacking order phase (LPSO) from kink bands (grain size control of the α-Mg phase) producing a fine dispersion of the LPSO phase in the α-Mg phases.

During last decade, a substantial advance is made in welding and joining of magnesium alloys. To improve in conventional fusion welding techniques, new methods and their hybrids are developed.

Confident that all these high-level contributions are of interest, I would like to extend my sincere acknowledgments to all those who contributed to the drafting of this book.

Prof. Dr. Waldemar Alfredo Monteiro
[1]School of Engineering - Presbyterian Mackenzie University,
[2]Materials Science and Technology Center –
Energetic and Nuclear Researches Institute São Paulo, S.P.,
Brazil

Section 1

Properties and Microstructure of Magnesium-Based Quasicrystals

Chapter 1

Mg-Based Quasicrystals

Zhifeng Wang and Weimin Zhao

Additional information is available at the end of the chapter

http://dx.doi.org/10.5772/48163

1. Introduction

Quasicrystals (QCs) are a well-defined ordered phase of solid matter with long-range quasiperiodic translational order and an orientational order [1], but no three dimensional translational periodicity [2]. In 1984, Shechtman et al [3] first reported these structures in a rapidly solidified Al–Mn alloy. It brings about a paradigm shift in solid-state physics for these atomic arrangements are forbidden for conventional crystallography [4] and have long been thought forbidden in nature. The unexpected discovery of QCs presents scientists with a new, puzzling class of materials and involves hundreds of researchers in this realm. During the beginning period for QC study, many QCs were fabricated in Al-based alloys [5]. Luo et al [6] discovered first Mg-based QCs in Mg-Zn-(Y, RE) system in 1993 which extend the alloy system of QCs.

So far, QCs in various systems have been synthesized in laboratories [2] and have also been discovered in a natural mineral [7] which comes from extraterrestrials. Many noticeable results were disclosed. The reported evidence [8] indicates that QCs can form naturally under astrophysical conditions and remain stable over cosmic timescales, giving unique insights on their existence in nature and stability. In 2011, the Nobel Prize in Chemistry was awarded to Daniel Shechtman for "the discovery of quasicrystals". Nowadays, scientists all over the world refocus these amazing materials and their promising applications.

As is well-known, QCs possess a host of unusual mechanical and physical properties [9] such as high strength, high thermal conductivity, and low friction coefficient [10]. Though they cannot be applied directly as structural materials for their innate brittleness, they can be used as good strengthening phases for some flexible matrix. Moreover, QCs have good corrosion resistance and were introduced into compounds which have been applied in some medical fields [11,12]. In this chapter, QC morphology evolution, its influence factors, QC-strengthened alloys and QC corrosion resistance are discussed. These basic researches are very useful for further development of QCs.

2. Morphologies of quasicrystals

QCs present fascinating three dimensional morphologies such as dodecahedral and icosahedral shapes (Fig.1). In different alloy systems, QC can be produced by slow-cooling method or rapidly solidified method. Mg-Zn-Y QCs possess a broad QC forming range. They can be synthesized in a common casting process [10].

(a) (b)

Figure 1. Fascinating quasicrystals [13] (a) Dodecahedral Zn-Mg-Ho single QC grain (b) Icosahedral Al-Mn QC flowers

2.1. Morphology evolutions of Mg-Zn-Y quasicrystals [14]

The $Mg_{72}Zn_{26.5}Y_{1.5}$ (at.%) alloys were produced by a reformed crucible electric resistance furnace (SG_2-5-10A, as shown in Fig.2), melted under the mixture of SF_6/CO_2 protective atmosphere. Stirring for 2 min by impellor at 1073K and holding for 5 min above 1053K, the melt was poured and cooled by different cooling media (as shown in Fig.3 and Table 1). The cooling curves (as shown in Fig.4) of the alloys were monitored by multichannel data acquisition cards. The results showed that, the cooling rate was sequentially decreased from cooling media 1 to 5. The SEM images of Alloy 1 ~ Alloy 5 were shown in Fig.5.

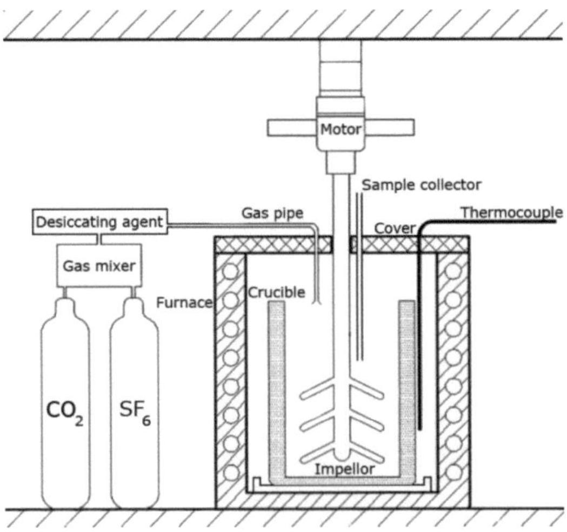

Figure 2. Schematic diagram of apparatus for making QC alloys

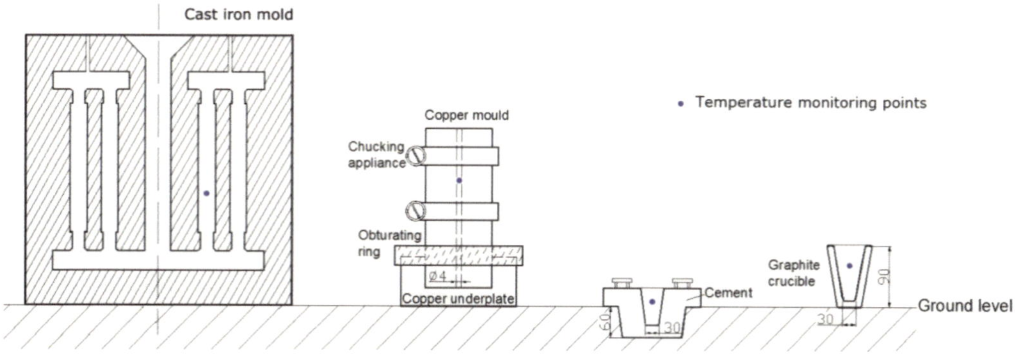

Figure 3. Schematic diagram of cooling media

Alloy no.	Cooling media
1	Be extracted by sample collector and cooled in water
2	Copper mould
3	Cast iron mould
4	Cement mould
5	Be poured into a graphite crucible and cooled in air

Table 1. Cooling media of the alloys

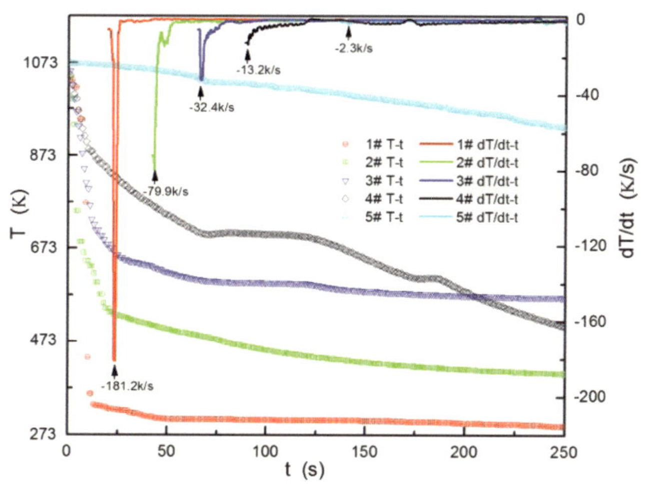

Figure 4. Cooling curves of the Alloys

The QC size gradually increased and the QC morphology changed with decreasing cooling rate. Decahedral quasicrystals (DQCs) were formed in Alloy 1 under cooling media 1, while icosahedral quasicrystals (IQCs) were formed in Alloy 2 ~ Alloy 5 under other cooling media. Moreover, the microhardness was larger for the smaller-sized QCs (Table 2). IQCs are quasiperiodic in three dimensions, while DQCs are quasiperiodic in two dimensions [2]. The DQCs formed in Alloy 1 presented flat bacilliform morphology and 10-fold symmetry characteristic. With decreasing cooling rate, the IQCs in Alloy 2 and Alloy 3 exhibited petal-like morphology under metal mould casting condition. Furthermore, the slower cooling rate induced larger IQC petals. With the further decrease of the cooling rate, the IQC petals showed nearly circular morphology. Finally, the IQCs grew up to large polygons in the slow cooling conditions.

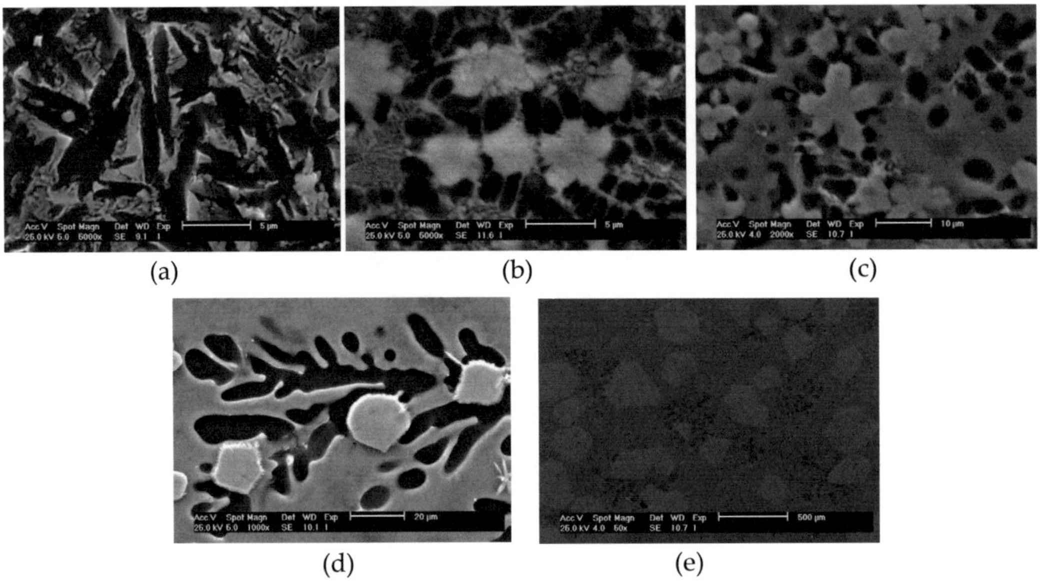

Figure 5. SEM images of Alloy 1~5 a) Alloy 1 (b) Alloy 2 (c) Alloy 3 (d) Alloy 4 (e) Alloy 5

Alloy no.	QC size / μm	QC morphology	QC microhardness / HV
1	10~12 in length	Flat X-shape	287
2	4~6	Petal-like	272
3	10~15	Petal-like	157
4	18~22	Near circular petal-like	182
5	300~400	Polygon	195

Table 2. Comparisons of the quasicrystals

In order to clarify how the IQCs transformed from morphology of Alloy 1 to Alloy 2, the $Mg_{72}Zn_{26}Y_{1.5}Cu_{0.5}$ alloys were synthesized under a water-cooled copper mold with pouring gate diameter of 2 mm and 4 mm. Such cooling rates were just between the cooling media 1 and 2. The cooling rate of water-cooled copper mold with pouring gate diameter of 2mm was faster than that of 4mm. Flat DQCs like Alloy 1, and spherical IQCs were formed respectively in Fig.6 (a) and (b), and pouring gate diameter was 2mm and 4mm correspondingly. We can see from Fig.6, a plane branch grew out in one of two-dimensional (2D) prior growth directions of the flat DQCs (marked by a red arrow in Fig.6 (a)). And then more branches grew out in three-dimensional (3D) directions (marked by a red arrow in Fig.6 (b)).

These branches increasingly became dense and agglomerate, and finally created a cluster for the primary IQC morphology.

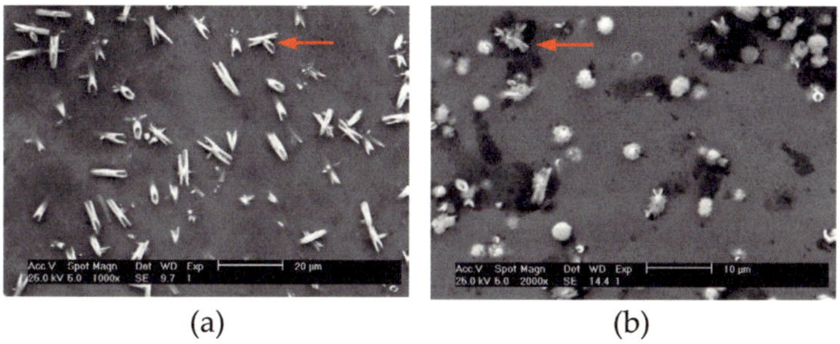

Figure 6. SEM images of $Mg_{72}Zn_{26}Y_{1.5}Cu_{0.5}$ alloys (a) Flat DQC (b) Spherical IQC

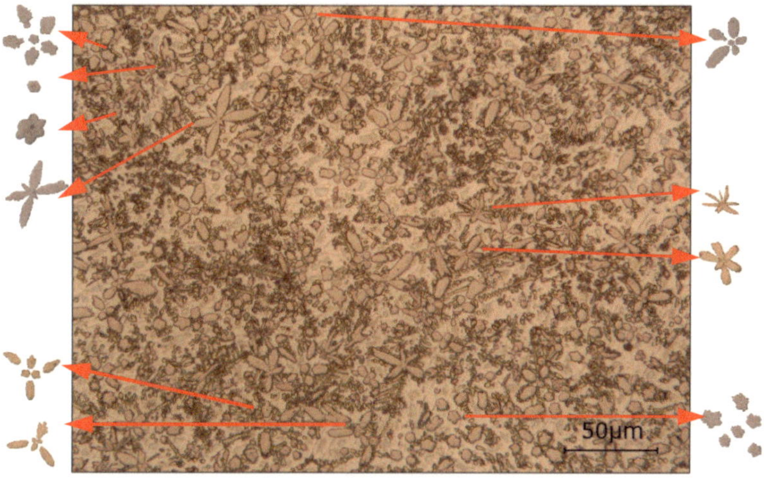

Figure 7. Optical microstructure of Alloy 3 after heat treatment at 750 K for 15 min

A heat treatment for Alloy 3 at 750 K for 15 min was prepared for studying IQC growth process between IQC morphology in Alloy 3 and in Alloy 4. It can be seen from Fig.7 that various shapes of QCs at different growth stages were formed in the heat treatment process. There were plentiful IQC nuclei in as-cast Alloy 3, but the growth was not complete due to a fast cooling process. The petals shown in Alloy 3 were the ones who had experienced the nucleation process only, but do not have enough time to grow up into the morphology in Alloy 4. During the heat treatment, the IQC nuclei continued to grow.

From the above, the IQC morphology evolution process between IQCs in Alloy 1 and Alloy 2 as well as between IQCs in Alloy 3 and Alloy 4 were revealed. A general drawing of morphology evolution of Mg-Zn-Y quasicrystal phase in growth process was shown in Fig.8. Twenty-two kinds of typical morphology of Mg-Zn-Y QC phase during cooling process were extracted from SEM and OM images.

During cooling process of Mg-Zn-Y alloys, at first a plane branch (shape 2) grew out in one of prior growth directions of the flat DQCs (shape 1). And then more branches emerged and created a cluster (shape 3), which was the primary morphology of IQCs. At the beginning of the IQC growth stage, its morphology was near spherical (shape 4). The spherical interface was not maintained with alteration of the ambience conditions. Along the prior growth directions, the spherical IQC sprouted five petals (shape 5) or six petals (shape 16). These petals subsequently grew up and became larger in length (shape 6 and shape 17), and further separated from each other (shape 8 and shape 18). The separated IQC petals grew up (shape 9) and became new independent IQCs (as shape 5). If there were still leftover Zn and Y elements in the melt, the IQC petals will continue to split and repeat the cycle from shape 5 to shape 9 until they were used up. With decrease of the cooling rate and increase of the growth time, the IQCs became maturity and grew bigger (shape 11), and finally grew into bulk polygons.

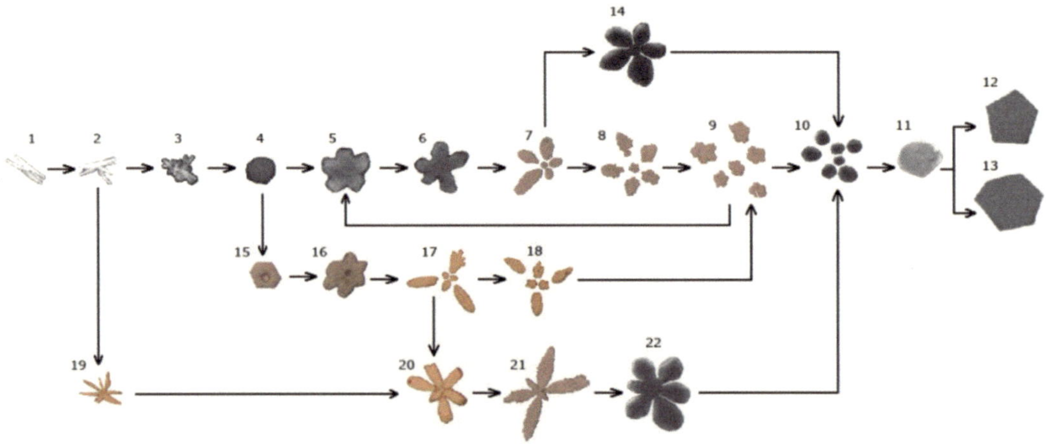

Figure 8. Schematic diagram of morphology evolution of Mg-Zn-Y quasicrystal phase in growth process

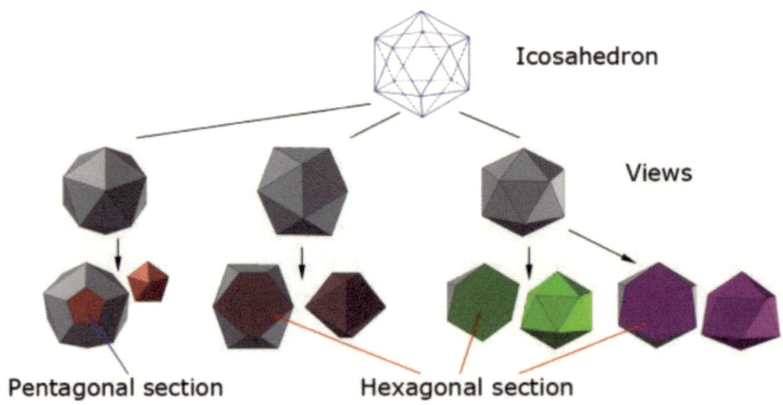

Figure 9. Section schematic diagram of icosahedrons

The reason why the final morphology of IQCs was pentagonal (shape 12) and hexagonal (shape 13) polygon can be showed in Fig.9: a mature Mg-Zn-Y quasicrystal is an icosahedron in a 3-D view; when we observe it in different directions, it show different views; and when we grind and polish samples in parallel direction to the views, pentagonal and hexagonal cross-section morphology are presented with multiple probability.

2.2. Effects of the fourth component and undercooling [15,16]

The solidified process of quasicrystal phases which consist of grain nucleation and subsequent growth is similar to crystals. It was necessary to properly control the cooling rate during these two processes for the formation of the quasicrystal phase is thermodynamically unstable. Lower cooling rate might not effectively suppress the crystallization and would result in the formation of crystal phase while higher cooling rate might suppress the nucleation and growth of the quasicrystal phase and would result in the formation of amorphous phase. For quasicrystal containing magnesium alloys, stable icosahedral quasicrystal phase (I-phase) can be obtained under normal casting conditions.

At the early stage of nucleation process, the single fourth component particles act as potential nucleating substrates, and the morphology of I-phase should be nearly spherical. Because the coalescence of the fourth component at solidification front, surface energy at that local region was elevated, and growing velocity of I-phase slowed down. Moreover, the same heat dissipating condition in all directions leads to the same growing velocity of I-phase in all directions. Furthermore, during this process, highest volume percentage of surface layer to the whole volume of I-phase particle resulted in highest surface energy of I-phase, which enabled the morphology of I-phase particle shrinking to spherical or near-spherical. Therefore, the solidified morphology of I-phase depended on the stability of spherical I-phase during the subsequent growth [17]. I-phase with spherical morphology would be obtained if I-phase forming initially could preserve spherical interface stable in the whole growth process. Otherwise, I-phase with irregular or dendrite morphology would be eventually generated. According to the research results of Mullins et al. [18], relative stability criterion of spherical interface with radius being R_r can be expressed by the rate of change per unit perturbation amplitude:

$$\frac{\dot{\delta}}{\delta} \leq \frac{\dot{R}}{R} \tag{1}$$

$$\frac{\dot{\delta}}{\delta} = \frac{(l-1)K_l}{K_s L}\left[\Delta T - \Gamma\frac{T_m \Gamma}{R}(1+\partial_t)\right] \tag{2}$$

The critical radius maintaining the spherical I-phase interface relative stable was:

$$R_r = \frac{2T_m \Gamma}{\Delta T}\left[1 + \frac{(l-1)}{(l-2)}\partial_t\right] \tag{3}$$

$$\partial_t = \frac{1}{2}(l+2)\left[1 + l\left(1 + \frac{K_s}{K_l}\right)\right] \tag{4}$$

Where δ is the amplitude of fluctuation, K_s the thermal conductivity of the solid phase, K_l the thermal conductivity of the liquid phase, L is latent heat of freezing, ΔT is degree of undercooling in the melt, Γ the ratio of interface energy to latent heat of solid phase per unit volume, l the rank of pherical harmonic function, T_m is the melting point of the alloy.

It can be known from Eqs. (3)~(4) that decreasing ΔT or elevating the interface energy between the I-phase and the melt were beneficial to the stability of spherical interface. The addition of a certain amount of the fourth component not only provided potential nucleating sites for I-phase, but also purified the melt by removing oxygen and the fourth component with harmful impurity elements. The coalescence of the fourth component compounds at solid/liquid interface resulted in higher interface energy and higher value of Γ. Moreover, the addition of the fourth component promoted heterogeneous nucleation of I-phase, lowered the degree of undercooling ΔT and increased the critical radius R_r. Meanwhile, the same heat dissipating condition of the I-phase particle in all directions resulted in the same growing velocity of I-phase particle in all directions, enabling I-phase to keep spherical growing front and providing positive conditions for spherical growth of I-phase.

However, if superfluous addition of the fourth component, un-dissolved fourth component will discharge from the solid phase to solid/liquid interface and formed the fourth component solute transitional layer with certain thickness. Moreover, due to the increasingly enrichment of the fourth component compounds in front of the growing solid/liquid interface of I-phase particle, the degree of constitutional under-cooling increased, and ΔT increased as well.

$$\Delta T = \Delta T_h + \Delta T_c + \Delta T_k \tag{5}$$

Where ΔT_h is thermodynamics undercooling, ΔT_c the constitutional undercooling, and ΔT_k the kinetics undercooling. It means that ΔT is composed of three parts of ΔT_h, ΔT_c and ΔT_k.

Increased ΔT intensified the instability of spherical growing surface of I-phase particle. Then the I-phase turn to coarse, the spherical morphology will be wrecked and transform to petal-like.

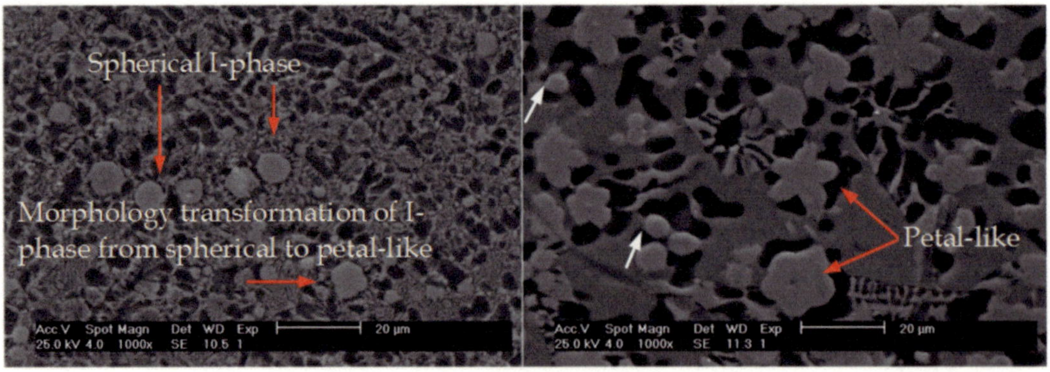

Figure 10. SEM images of as-cast Mg-Zn-Y-Sb alloys containing I-phase; (a) $Mg_{72.2}Zn_{26.2}Y_{1.5}Sb_{0.1}$ (b) $Mg_{72.1}Zn_{26.2}Y_{1.5}Sb_{0.2}$

Fig.10 shows SEM images of Mg-Zn-Y-Sb alloys. I-phase morphology in $Mg_{72.2}Zn_{26.2}Y_{1.5}Sb_{0.1}$ was spherical while $Mg_{72.1}Zn_{26.2}Y_{1.5}Sb_{0.2}$ presented petal-like. It can be seen from Fig.10(a) that the value of critical radius R_r of I-phase in Mg-Zn-Y-0.1Sb alloy was about 8μm when the content of the fourth component Sb was 0.1%. If local conditions changed, and spherical radius value exceed R_r, the morphology transformation of I-phase from spherical to petal-like will be occurred (marked by the lower red arrow in Fig.10(a)). So we can see that the superfluous addition of the fourth component was negative to the stability of spherical interface, and also made against to forming spherical I-phase. We can see from Fig.10(b): most parts of I-phase are petal-like while a few of I-phase are spherical(marked by white arrows). Therefore a critical stable radius indeed exists. Once the interface radius of I-phase is larger than R_r in IQC growth process, the final morphology of I-phase in that local zone will be petal-like. Conversely, spherical morphology will be preserved in local zone if the interface radius of I-phase is smaller than R_r.

The effect of different contents of the fourth component and different degree of undercooling on critical stable radius of spherical I-phase can be shown in Fig.11. As we discussed above, for certain cooling conditions and certain compositions of Mg-Zn-Y alloys, certain size of critical stable radius exist and we describe this state as state I. The addition of a small amount of the fourth component is able to result in an decrease of degree of undercooling and finally increase the critical stable radius R_r as seen in Eqs.(3). We can describe this state as state II. However, if superfluous addition of the fourth component, constitutional undercooling will come out, ΔT will increase. Thus the critical stable radius of spherical I-phase will decrease. This state can be called state III.

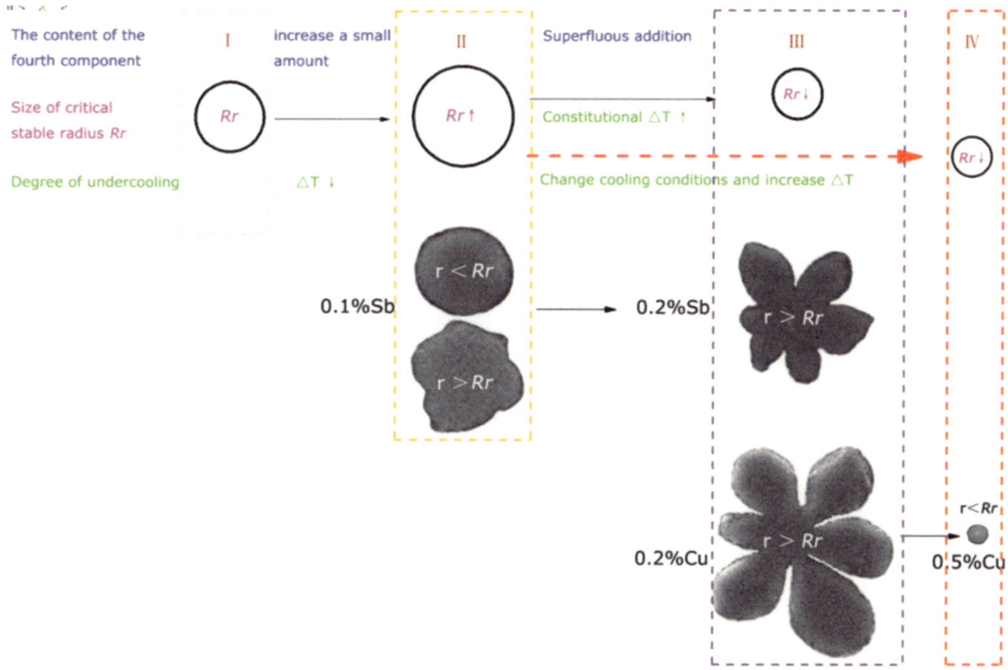

Figure 11. Schematic diagram of different states and transform process of critical stable radius

For a certain cooling condition and a certain composition of alloys, different contents of the fourth component and their critical stable radius have relationships of one-to-one correspondence. Fig.11 takes the fourth component Sb and Cu for examples. Only when the radius of IQC less than R_r in their respective states can spherical IQC be formed. Under most of the conditions, if superfluous addition of the fourth component, small-sized R_r will generate big-sized petal-like IQC. It seems as if superfluous addition of the fourth component could not produce spherical I-phase. Actually, we can improve cooling conditions and increase ΔT_h and ΔT artificially. Much smaller critical stable radius will make it difficult to forming spherical I-phase. However, higher cooling rate might cut down the growth time of the quasicrystal phase. Spherical interface of I-phase forming preliminary stage will be stably preserved in the whole growth process, and then smaller-sized spherical I-phase which its radius less than R_r will occurred. We can define this state as state IV. Under these principles, a kind of spherical I-phase with high content of the fourth component but amazing minisize (as shown in Fig.12(b)) can be produced by using a water-cooled copper mould (as shown in Fig.13(a)). So, it is a novel way to produce spherical I-phase with high content of the fourth component in minisize by increasing thermodynamics undercooling artificially. In this way, we can easily control cooling rate in a certain range and obtain quarternary spherical IQC with different minisize scale.

Searching proper content of different fourth component, confirming the size of spherical stable radius, developing quarternary spherical IQC with different minisize scale, and thoroughly making good use of IQC particles as reinforcement phase are future problems and proper research points.

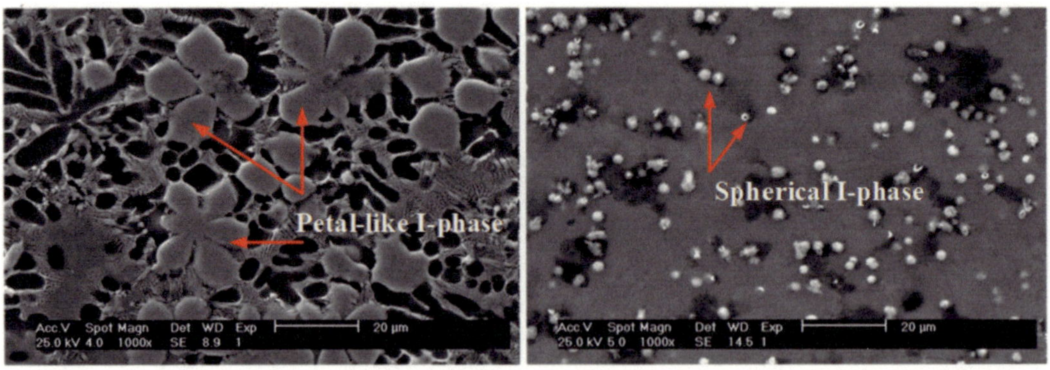

Figure 12. SEM images of Mg-Zn-Y-Cu alloys cooled in different mould(a) $Mg_{72.1}Zn_{26.2}Y_{1.5}Cu_{0.2}$ (cast iron mould) (b) $Mg_{72.0}Zn_{26.0}Y_{1.5}Cu_{0.5}$ (water-cooled Cu mould)

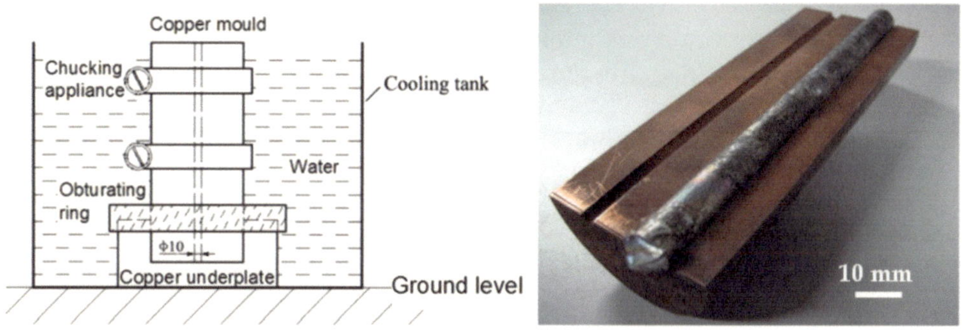

(a) (b)

Figure 13. Mould for produces spherical QC alloys; (a) Water-cooled cooper mould (b) Casting

3. Effects of quasicrystal alloys on mechanical properties of magnesium alloys [19]

The effects of different Ce contents on microstructure of Mg-Zn-Y-Ce QC alloys are shown in Fig.14. Mg-Zn-Y QCs showed petal-like morphology under cast iron mould cooling conditions. When the added Ce content was small (0.2 at.%), the morphology and size of QC petals were basically unchanged. With the increase of Ce content (0.5 at.%), the amounts and size of the QC petals were significantly increased, and the petals became more round. When the Ce content reached 0.8 at.%, the amounts of I-phases further multiplied, but the petals reduced in size. The petal branch became short, unconspicuous, and subsphaeroidal. With

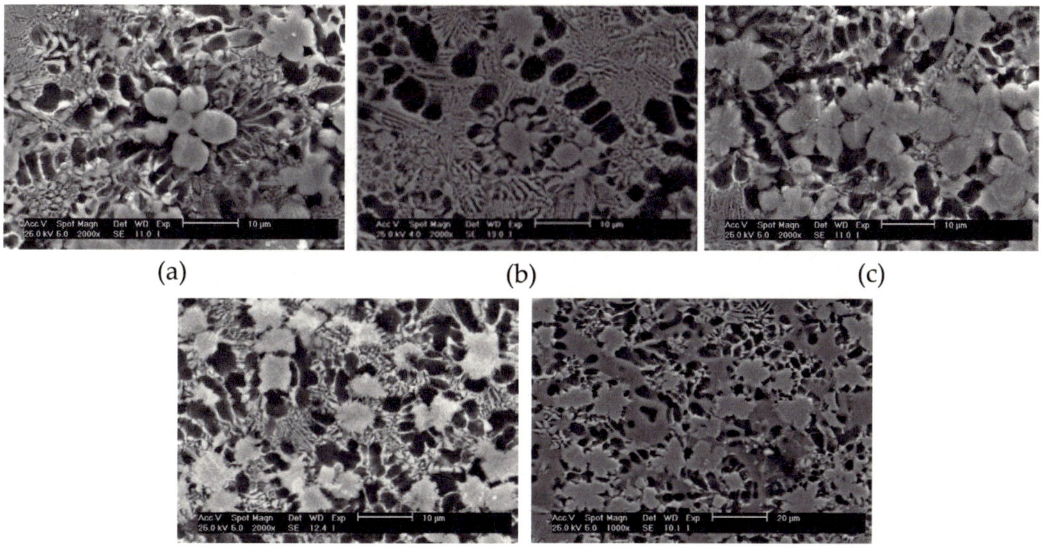

(a) (b) (c) (d) (e)

Figure 14. SEM images of the Mg-Zn-Y-Ce QC alloys. (a) $Mg_{72.5}Zn_{26}Y_{1.5}$ (b) $Mg_{72.3}Zn_{26}Y_{1.5}Ce_{0.2}$ (c) $Mg_{72.1}Zn_{25.9}Y_{1.5}Ce_{0.5}$ (d) $Mg_{72}Zn_{25.7}Y_{1.5}Ce_{0.8}$ (e) $Mg_{71.8}Zn_{25.7}Y_{1.5}Ce_{1.0}$

the further increase in Ce content (1.0 at.%), the IQC petal size grew twice that of 0.5 at.% Ce, and they were transformed as multi-secondary dendrites of the five- or six-petaled flowers. This process was in line with the cooling influencing law [15].

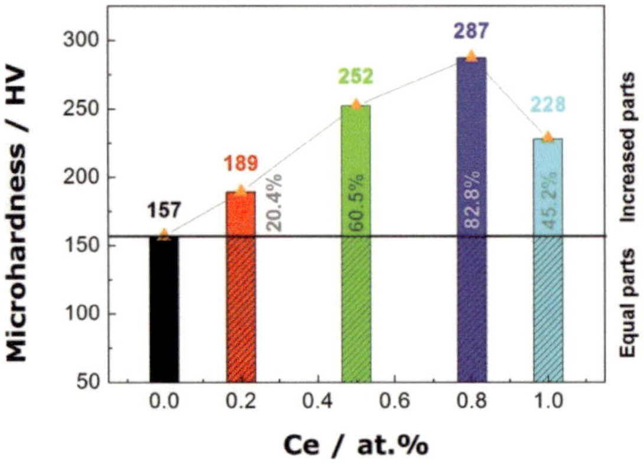

Figure 15. Microhardness of quasicrystals.

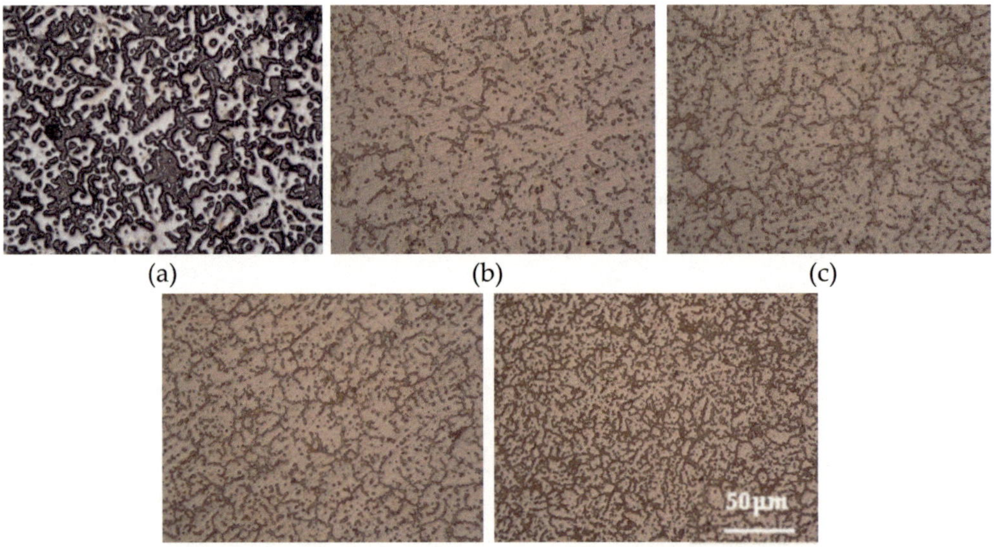

(a) (b) (c)

(d) (e)

Figure 16. Microstructure of AZ91 alloys reinforced by different content of $Mg_{72}Zn_{25.7}Y_{1.5}Ce_{0.8}$ alloys (wt%). (a) 0% (b) 5% (c) 10% (d) 15% (e) 30%

The microhardness test results (as shown in Fig.15) of IQC alloys showed as the following: All values of microhardness of quaternary QCs were higher than those of ternary QCs. With increase in Ce content, the microhardness of I-phase also increased. However, when the dosage reached a certain value (i.e., 1.0%), the microhardness of I-phase decreased sharply. The microhardness value of I-phase in the $Mg_{72}Zn_{25.7}Y_{1.5}Ce_{0.8}$ alloy reached to HV287, which is 82.8% higher than that in ternary $Mg_{72.5}Zn_{26}Y_{1.5}$ alloy. In following experiments,

$Mg_{72}Zn_{25.7}Y_{1.5}Ce_{0.8}$ alloy was used as a master alloy to strengthen AZ91 alloys since the quaternary subsphaeroidal I-phase contain high microhardness and possess better wetting power with Mg matrix.

$Mg_{72}Zn_{25.7}Y_{1.5}Ce_{0.8}$ master alloys with contents of 0%, 5%, 10%, 15%, and 30% (wt.%) were added into AZ91 alloys. Changes in the microstructure of AZ91 alloys are shown in Fig. 16. With the increase in the amount of the $Mg_{72}Zn_{25.7}Y_{1.5}Ce_{0.8}$ alloy, the grains of AZ91 alloys were gradually refined, while β-phase was refined and narrowed. However, when the dosage of $Mg_{72}Zn_{25.7}Y_{1.5}Ce_{0.8}$ alloy was too high (30%), β-phase turn to coarse.

In this craft, $Mg_{72}Zn_{25.7}Y_{1.5}Ce_{0.8}$ alloy was added into molten AZ91 and remelted. In the subsequent metal mold cooling process, the I-phases nucleated, but insufficient time did not allow for the adequate increase in size. Therefore, small granular I-phases precipitated from the grain interiors of the AZ91 alloys. These granular I-phases mixed with divorced β-phase particles, which baffled the process of identification of one from the other. In several kinds of phases of $Mg_{72}Zn_{25.7}Y_{1.5}Ce_{0.8}$ alloy, only I-phases remained after remelting. Other phases integrated into the AZ91 and became constituting elements of AZ91 alloys. Since I-phases are heat-stable phases [20], they remain in the alloys and will not be broken down into other phases even in high-temperature heating process. Thus, they can play significant roles for the matrix after heat treatment. Considering this characteristic of I-phases, we can study the effects of heat treatment to further improve on the mechanical properties of QCs reinforced AZ91 alloys.

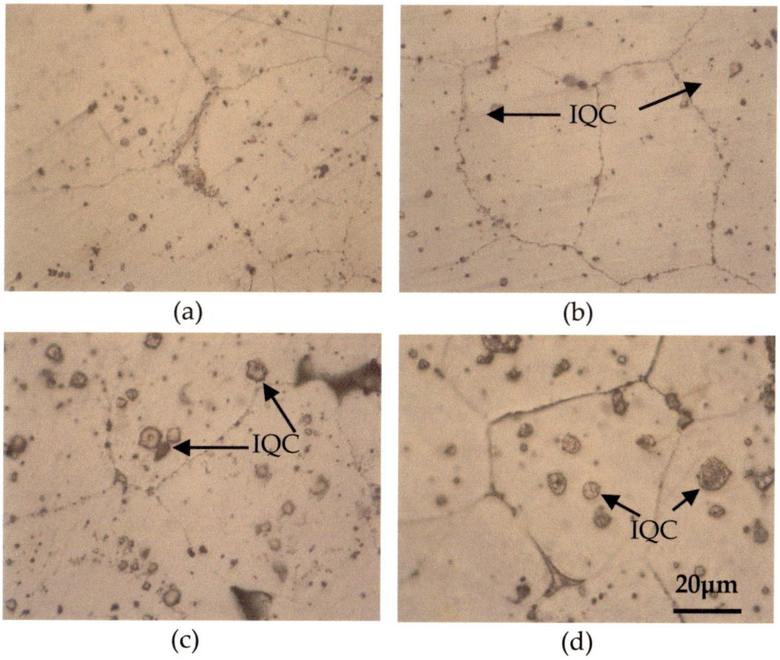

Figure 17. Microstructure of AZ91 alloys reinforced by different content of $Mg_{72}Zn_{25.7}Y_{1.5}Ce_{0.8}$ alloys after T4 solution treatment at 420°C for 24h. (wt%) (a) 5% (b) 10% (c) 15% (d) 30%

After solution treatment (420°Cx24h), grain boundaries of AZ91 alloys became clear, the typical reticular morphology of β-phase disappeared, and I-phases and Al-Mn particles precipitated in the intragranular zone. It was difficult to distinguish between the two particles when the content of $Mg_{72}Zn_{25.7}Y_{1.5}Ce_{0.8}$ master alloy was low. I-phase was formed through the reaction of L→α-Mg+I at about 400°C during solidification process [10]. Therefore, under this temperature, small IQC particles increased in size and ripened during the long time process of T4 heat treatment. As shown in Fig.17, during the same heat treatment process, with the increase of $Mg_{72}Zn_{25.7}Y_{1.5}Ce_{0.8}$ alloy, the amounts and size of quaternary Mg-Zn-Y-Ce IQCs in AZ91 matrix gradually increased. The Al-Mn phases, however, did not change to bigger. This made the two kinds of particles distinguishable.

An aging treatment (220°Cx8h) was conducted after the solution treatment. With an aging temperature of 220°C set between the continuous precipitation temperature (310°C) and discontinuous precipitation temperature (150°C), but nearer to the discontinuous precipitation temperature, the β-phases of AZ91 alloys mainly discontinuously precipitated. During the 8h aging treatment process, lamellar precipitates formed from the grain boundaries and grew in the intragranular. Granular β-phase also precipitated in the intragranular through a continuous precipitation method. Thus, precipitates filled the whole grain, as shown in Fig. 18.

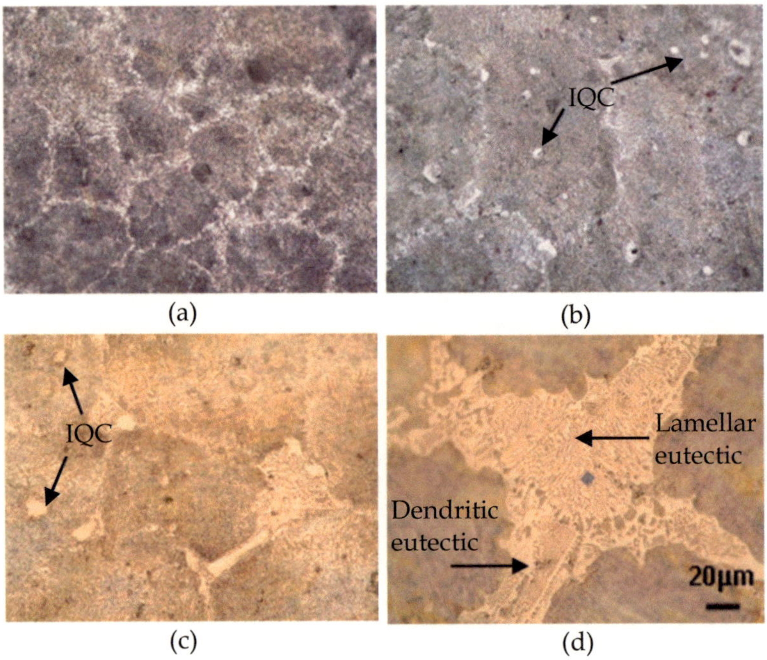

Figure 18. Microstructure of AZ91 alloys reinforced by different content of $Mg_{72}Zn_{25.7}Y_{1.5}Ce_{0.8}$ alloys after T6 solution(420°Cx24h) and aging(220°Cx8h) treatment. (wt%) (a) 5% (b) 10% (a) 15% (b) 30%

I-phase was difficult to be observed after the aging treatment when the content of $Mg_{72}Zn_{25.7}Y_{1.5}Ce_{0.8}$ alloy was small (5%). With an increase in the content of $Mg_{72}Zn_{25.7}Y_{1.5}Ce_{0.8}$

alloy, the amounts of IQCs in the grain of AZ91 alloys likewise increased. When the content of $Mg_{72}Zn_{25.7}Y_{1.5}Ce_{0.8}$ alloy continued to rise, the size of IQCs turned larger, but eutectic phases in grain boundaries became coarse. With the excessive addition of $Mg_{72}Zn_{25.7}Y_{1.5}Ce_{0.8}$ alloy, only a few I-phases remained in the intragranular AZ91 alloys; eutectic phases in the grain boundary became very thick, and the morphology of eutectic β-phase presented a lamellar. Meanwhile, parts of the eutectic α-Mg showed dendrite morphology.

Fig.19 shows that the value of the Brinell hardness (HB) of the IQC-reinforced AZ91 alloy decreased after the solution treatment, while its value remarkably increased after the further aging treatment. With the increasing addition of $Mg_{72}Zn_{25.7}Y_{1.5}Ce_{0.8}$ alloy, the HB values of as-cast and solution-treated AZ91 alloys showed a linear increase, while the HB values of aging-treated AZ91 alloys first increased and then decreased.

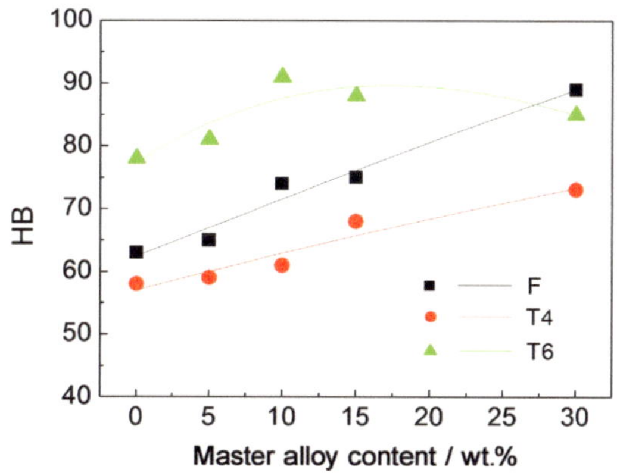

Figure 19. Relationship between additions of $Mg_{72}Zn_{25.7}Y_{1.5}Ce_{0.8}$ master alloy and Brinell hardness of AZ91 alloys.

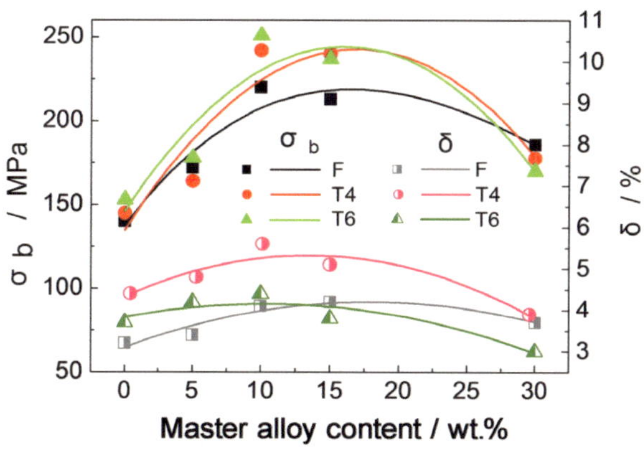

Figure 20. Relationship between additions of $Mg_{72}Zn_{25.7}Y_{1.5}Ce_{0.8}$ master alloy and mechanical properties of AZ91 alloys.

Fig.20 shows that the values of tensile strength (σ_b) and elongation (δ) of AZ91 alloys with all states reached their maximum when the content of $Mg_{72}Zn_{25.7}Y_{1.5}Ce_{0.8}$ alloy was about

10%. With increasing content of $Mg_{72}Zn_{25.7}Y_{1.5}Ce_{0.8}$ alloy, the mechanical properties of AZ91 alloys increased first and decreased subsequently.

After adding $Mg_{72}Zn_{25.7}Y_{1.5}Ce_{0.8}$ alloy into AZ91 alloys, the introduced Y and Ce elements played mixed roles in grain refinement and strengthening. Tensile strength and elongation of AZ91 alloys increased. Furthermore, a large number of introduced highly hardened IQC particles shifted the HB value of as-cast AZ91 and the value increased with the rising content of $Mg_{72}Zn_{25.7}Y_{1.5}Ce_{0.8}$ master alloy. The excessive addition of $Mg_{72}Zn_{25.7}Y_{1.5}Ce_{0.8}$ alloy reduced the mechanical properties of AZ91 alloys; these were related to the formation of coarse β-phase, which produced dissevered effects to the matrix in the deformation process.

After the solution treatment, the majority of the main strengthening phase (reticulated β-phase) of AZ91 alloys disappeared, which made the HB value of solution-treated AZ91 alloys lower than in the as-cast. In addition, the microstructure of AZ91 alloys became homogeneous due to the annealing treatment. This eliminated most of the stress concentration and composition segregation. As a result, the tensile properties and plasticity of the heat-treated state AZ91 alloys showed small improvements compared to the as-cast AZ91 alloys. With additions of $Mg_{72}Zn_{25.7}Y_{1.5}Ce_{0.8}$ master alloy exceeding 10%, the reduced mechanical properties of AZ91 alloys resulted to large I-phases and dissevered effects to the matrix in the deformation process.

After the aging treatment, the lamellar eutectic β-phases that grew on the grain boundaries were parallel or perpendicular to the matrix; this played an important role in its strengthening. Due to the discontinuous precipitation of lamellar β-phases, with their main strengthening effect coming from this kind of precipitation method, in addition to continuous precipitation of pellet β-phases, the values of HB and tensile strength of AZ91 alloys rapidly increased. However, with the large amount of $Mg_{72}Zn_{25.7}Y_{1.5}Ce_{0.8}$ master alloy, the excess introduced a Y element, which brought about highly stable Al-Y phases during the aging treatment. These Al-Y phases resulted to a pinning effect on the nucleation and growth of β-phases, thereby preventing the precipitation of β-phases. Thus, the β-phases on the grain boundaries were very coarse and did not grow in the intragranular zone (as shown in Fig.18(d)). Thick and hard β-phases can easily make cutting effects to the matrix. Their interfaces can easily be crack sources of the AZ91 alloys, which is unfavorable to the strength and plasticity of magnesium alloys. As a result, the tensile strength and elongation of AZ91 alloys decreased sharply.

4. Mg-based nano-quasicrystals [21,22]

In previous study [14~16, 19, 21~27], the effects of cooling conditions, heat treatment and the fourth components on QC morphology, size and volume fractions are detailedly researched. Spherical QCs with small size are fabricated in a relatively high cooling rate. In this part, we improve the cooling condition by using a water-cooled wedge-shaped copper mould (Fig. 21 shows its casting) to produce QCs in nanoscale.

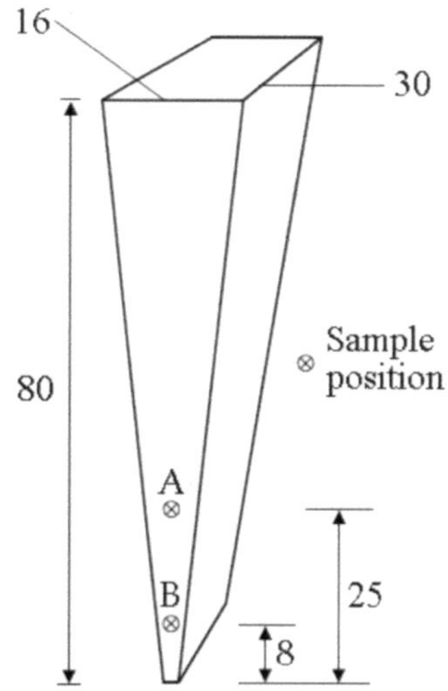

Figure 21. Sketch map of a wedge-shaped casting (mm)

TEM photos of QC alloys (Alloy compositions are listed in Table 3) in different sample positions are shown in Fig. 22. Three kinds of componential micro/nano QC phases are synthesized on tip of wedge-shaped castings. Energy-dispersive spectroscopy (EDS) analysis (Fig. 23) shows that micro/nano QC phases in Position B of Alloy 6 ~ Alloy 8 are Mg-Zn-Y phase, Mg-Zn-Y-Cu phase and Mg-Zn-Y-Cu-Ni phase, respectively. The selected area electron diffraction (SAED) patterns with typical five-fold rotational symmetry identify that these micro/nano QC phases are icosahedral QCs.

Alloy No.	Alloy compositions (at. %)				
	Mg	Zn	Y	Cu	Ni
6	72.0	26.0	2.0	-	-
7	71.0	26.0	2.0	1.0	-
8	71.0	26.0	2.0	0.5	0.5

Table 3. Nominal composition of the experimental alloys

Alloy No.	Sample position	QC size(diameter) / nm	QC morphology	Microhardness / HV
6	A	650-900	Petal-like	324
	B	330-340	Spherical	375
7	A	370-400	Spherical	367
	B	1.0-5.0	Spherical	459
8	A	20-55	Spherical	412
	B	8-30	Spherical	438

Table 4. Comparisons of the quasicrystals (QCs)

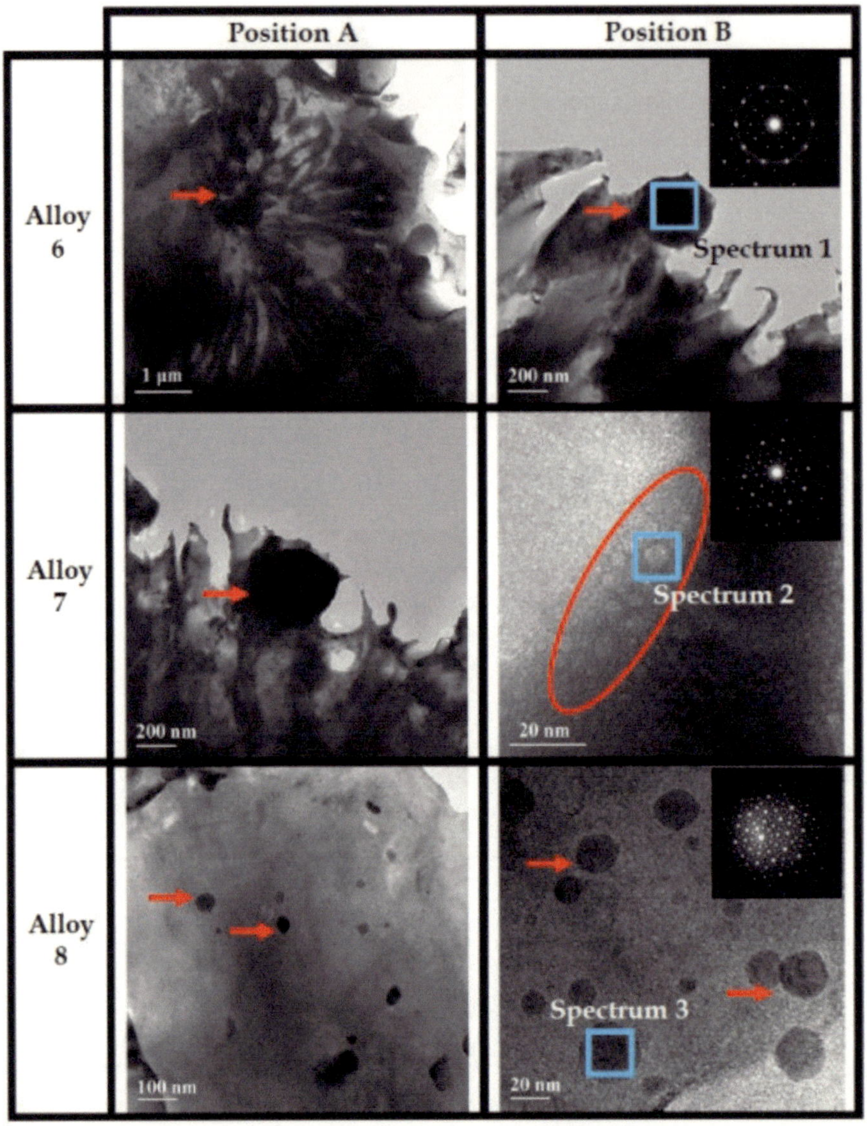

Figure 22. TEM photos of micro/nano-QC alloys and QC typical selected area electron diffraction (SAED) patterns on Position B of different alloys

Among all QCs, QCs in Position A of Alloy 6 show petal-like morphology, while others show spherical morphology. From the further analysis in Table 4, we can see that in alloys with same components, QCs in Position B are smaller than those in Position A, while QC microhardness in Position B is greater than that in Position A. After introducing Cu(-Ni) into Mg-Zn-Y alloys, we can see in the same sample position, QC size of Alloy 7 and Alloy 8 is obviously smaller than that of Alloy 6. QC size of Alloy 7 in Position A is close to that of Alloy 6 in Position B. Nano-QC spheres about 8~30 nm and 1~5 nm are synthesized in Position B of Alloy 8 and Alloy 7, respectively. It shows from the microhardness testing that the smaller the QC spheres, the greater their value of microhardness. Furthermore, the microhardness of nano-QC spheres in Position B of Alloy 7 exceeds HV450 which show fascinating properties.

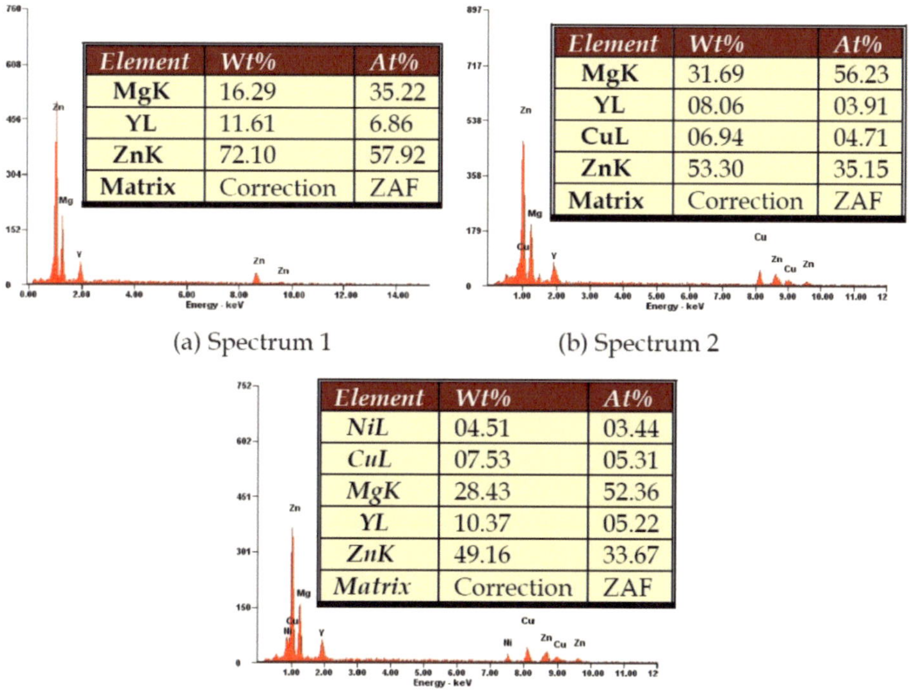

Element	Wt%	At%
MgK	16.29	35.22
YL	11.61	6.86
ZnK	72.10	57.92
Matrix	Correction	ZAF

(a) Spectrum 1

Element	Wt%	At%
MgK	31.69	56.23
YL	08.06	03.91
CuL	06.94	04.71
ZnK	53.30	35.15
Matrix	Correction	ZAF

(b) Spectrum 2

Element	Wt%	At%
NiL	04.51	03.44
CuL	07.53	05.31
MgK	28.43	52.36
YL	10.37	05.22
ZnK	49.16	33.67
Matrix	Correction	ZAF

(c) Spectrum 3

Figure 23. Energy-dispersive spectroscopy (EDS) analysis on QCs in Position B

Fig. 24 shows the potentiodynamic polarization curves of QC alloys (Position B) measured in simulated seawater open to air at room temperature. We can see that $Mg_{71}Zn_{26}Y_2Cu_1$ nano-QC alloy presents high corrosion resistance in simulated seawater and its corrosion resistance is much better than that of $Mg_{72}Zn_{26}Y_2$ and $Mg_{71}Zn_{26}Y_2Cu_{0.5}Ni_{0.5}$ QC alloys. The further study shows that this result can be ascribed to the existence of well-distributed nano-QC phases (shown in Fig. 25 by red arrows) and polygonal $Mg_2(Cu,Y)$ phases [28]. These high corrosion resistance phases decrease the anodic passive current density, improve the polarization resistance, cut down the corrosion rate (Table 5) and finally improve the corrosion resistance of the Mg-Zn-Y-based alloy markedly. Cu and Ni have long been considered as harmful elements for improving corrosion resistance of Mg-based alloy [29], however, they are used to synthesize nano-QC spheres in this paper. Due to high corrosion resistance of QC phases, $Mg_{71}Zn_{26}Y_2Cu_1$ and $Mg_{71}Zn_{26}Y_2Cu_{0.5}Ni_{0.5}$ nano-QC alloys present better corrosion resistance than $Mg_{72}Zn_{26}Y_2$ QC alloy. Moreover, the corrosion resistance of $Mg_{71}Zn_{26}Y_2Cu_1$ nano-QC alloys is higher than $Mg_{71}Zn_{26}Y_2Cu_{0.5}Ni_{0.5}$ nano-QC alloys for the higher damage level of Ni to the corrosion resistance of magnesium alloy than that of Cu when they have same contents [29].

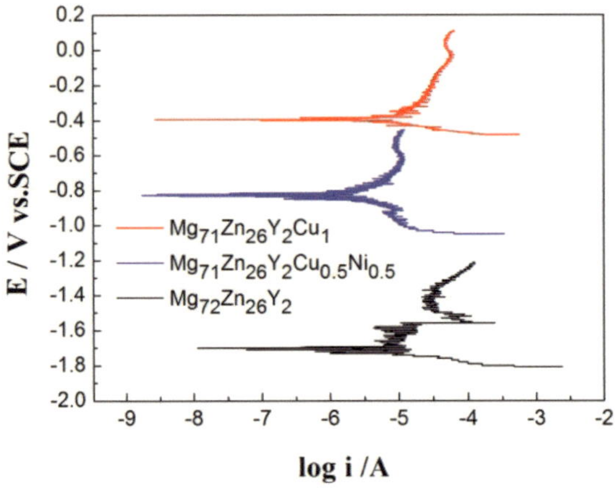

Figure 24. Potentiodynamic polarization curves of QC alloys (Position B) measured in simulated seawater open to air at room temperature

It was reported that a large negative enthalpy of mixing and/or existence of Frank-Kasper-type phases appear to be the crucial criteria for the formation of nanoquasicrystalline phase in any system [30]. Meanwhile, Mg-Zn-Y-based QCs just belong to Frank-Kasper-type phases [31] and have a certain negative enthalpy of mixing. So theoretically, Mg-Zn-Y-based nano-QCs can be formed in a proper cooling condition. The past cooling rate the researchers made to produce QCs was whether too high or too low, and was not content with the forming conditions of nano-QCs. This route just meets the demands for forming nanoscale QCs. So, nano-QCs are successfully produced in this paper. Moreover, the additions of Cu and Ni improve the degree of constitutional supercooling of Mg-Zn-Y melts and reduce the crucial criteria radius for forming spherical QCs. However, increasing thermodynamics undercooling coming from water-cooled wedge-shaped copper mould make it still possible to form spherical QCs. At the same time, the alloy components designed for this study is based on the three empirical rules [32] for the formation of metallic glass. It has been widely accepted that quasicrystals and at least some metallic glasses are built up with icosahedral clusters [33]. The short-range atomic configuration is very similar between the quasicrystal and amorphous phases [34]. On the tip of the wedge-shaped ingots, its cooling conditions are just suitable for these icosahedral clusters to be nucleation of QCs. And then, it leaves very short time for quasicrystal growth. So, it is nano-QCs that form in this route instead of metallic glasses.

Specimen	Icorr/μA/cm²	Rp/kΩ	Corrosion Rate/mpy
6#	11.09	6.925	19.298
7#	2.035	14.76	1.522
8#	3.762	8.105	3.084

Table 5. Corrosion parameters obtained from potentiodynamic polarization curves for Position B of QC alloys in simulated seawater. Icorr: corrosion current; Rp: polarization resistance.

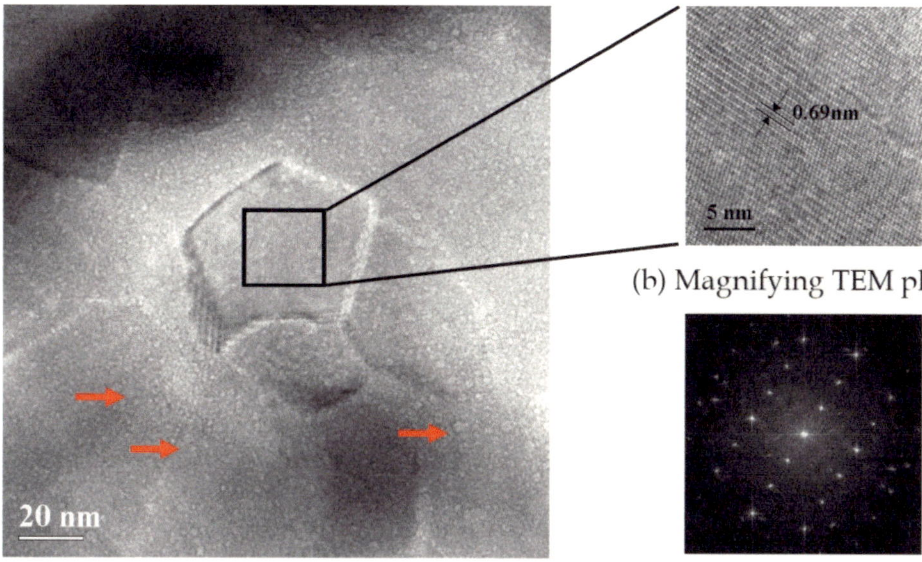

(a) TEM photo

(b) Magnifying TEM photo

(c) Fast Fourier transform (FFT) pattern

Figure 25. Pentagonal nanophase in $Mg_{71}Zn_{26}Y_2Cu_1$ alloy

5. Summary

The existing results show that QC characteristics are influenced by the cooling conditions during QC nucleation and subsequent growth. In macroscopic view, transformations of QCs in morphology, size and volume fractions are caused by changes of the cooling rate, the additions of fourth components and heat treatments. The further theoretical research shows that the final morphology of a QC is determined by the critical stable radius R_r. Only when the real radius of a QC less than R_r can spherical IQC be formed. Otherwise, petal-like QCs will form.

QC master alloys can be used to strengthen magnesium alloys. Proper doses may induce an improvement in mechanical properties of a magnesium alloy. Furthermore, we can fabricate nano-QCs by controlling thermodynamics undercooling and using a water-cooled wedge-shaped copper mould. Due to the good corrosion resistance of QCs, nano-QCs containing magnesium alloy show higher corrosion resistance.

Although QCs have been studied for about 30 years by scientists all over the world, successful applications of QCs have been very limited. For example, QCs can be applied as a surface coating for frying pans, could be used in surgical blades, and could be incorporated into hydrogen storage materials [2]. These are insufficient to meet people's demand for this amazing material. New applications are expected to develop.

Author details

Zhifeng Wang
School of Materials Science and Engineering, Hebei University of Technology, P. R. China

Weimin Zhao
Corresponding Author
School of Materials Science and Engineering, Hebei University of Technology, P. R. China

Acknowledgement

We are grateful for the financial support of the Natural Science Foundation of Hebei Province, China (No. E2010000057, No. E2010000121) and the International Science & Technology Cooperation Program of China (No. 2010DFA51850).

6. References

[1] Levine, D.; Steinhardt P.J. (1986). Quasicrystals. I. Definition and Structure. *Physical Review B*, vol.34, pp. 596-616, ISSN 1098-0121.

[2] Louzguine-Luzgin, D.V.; Inoue, A. (2008). Formation and Properties of Quasicrystals. *Annual Review of Materials Research*, Vol.38, pp. 403-423, ISSN 1531-7331.

[3] Shechtman, D.; Blech I.; Gratias D.; Cahn J.W. (1984). Metallic Phase with Long-range Orientational Order and No Translational Symmetry. *Physical Review Letters*, Vol.53, pp. 1951-1953, ISSN 0031-9007.

[4] Patricia, T. (2007). Quasicrystals – When All Pieces Fit Together. *Nature Materials*, Vol.6, pp. 11-12, ISSN 1476-1122.

[5] Tsai, A.P.; Inoue, A.; Masumoto, T. (1989). Stable Decagonal Al-Co-Ni and Al-Co-Cu Quasicrystals. *Materials Transactions*, Vol.30, pp. 463-473, ISSN 1345-9678.

[6] Luo, Z.P.; Zhang, S.Q.; Tang, Y.L.; Zhao, D.S. (1993). Quasicrystals in As-cast Mg-Zn-RE Alloys. *Scripta Metallurgica et Materialia*, Vol.28, pp. 1513-1518, ISSN 0956-716X.

[7] Bindi, L.; Steinhardt, P.J.; Yao, N.; Lu, P.J. (2009). Natural Quasicrystals. *Science*, Vol.324, pp. 1306-1309, ISSN 0036-8075.

[8] Bindi, L.; Eiler, J.M.; Guan, Y.B.; et al (2012). Evidence for the Extraterrestrial Origin of a Natural Quasicrystal. *Proceedings of the National Academy of Sciences of the United States of America (PNAS)*, Vol.109, pp. 1396-1401, ISSN 0027-8424.

[9] Blaaderen, A.V. (2009). Quasicrystals from Nanocrystals. *Nature*, Vol.461, pp. 892-893, ISSN 0028-0836.

[10] Zhang, Y.B; Yu, S.R.; Zhu, X.Y.; Luo, Y.R. (2008). Study on As-cast Microstructures and Solidification Process of Mg–Zn–Y Alloys. *Journal of Non-Crystalline Solids*, Vol.354, pp. 1564-1568, ISSN 0022-3093.

[11] Anderson, B.C.; Bloom, P.D.; Baikerikar, K.G.; et al (2002). Al–Cu–Fe Quasicrystal/Ultra-high Molecular Weight Polyethylene Composites as Biomaterials for Acetabular Cup Prosthetics. *Biomaterials*, Vol.23, pp. 1761-1768, ISSN 0142-9612.

[12] Schwartz, C.J.; Bahadur, S.; Mallapragada, S.K. (2007). Effect of Crosslinking and Pt–Zr Quasicrystal Fillers on the Mechanical Properties and Wear Resistance of UHMWPE for Use in Artificial Joints. *Wear*, Vol.263, pp. 1072-1080, ISSN 0043-1648.
[13] Abe, E.; Yan Y.; Pennycook, S.J. (2004). Quasicrystals as Cluster Aggregates. *Nature Materials*, Vol.3, pp. 759-767, ISSN 1476-1122.
[14] Wang, Z.F.; Zhao W.M.; Li, H.P.; et al. (2011). Morphologies of Quasicrystal Phase in Ternary Mg-Zn-Y Alloys. *Materials Science Forum*, Vol.675-677, pp. 215-218, ISSN 0255-5476.
[15] Wang, Z.F.; Zhao W.M.; Hur, B.Y.; et al. (2009). Effects of the Fourth Component and Undercooling on Morphology of Primary Mg-Zn-Y Icosahedral Quasicrystal Phase under Normal Casting Conditions. *China Foundry*, Vol.6, pp. 293-299, ISSN 1672-6421.
[16] Wang, Z.F.; Zhao W.M.; Li, H.P.; et al. (2010). Effect of Titanium, Antimony, Cerium and Carbon Nanotubes on the Morphology and Microhardness of Mg-based Icosahedral Quasicrystal Phase. *Journal of Materials Science & Technology*, Vol.26, pp. 27-32, ISSN 1005-0302.
[17] Zhang, J.S.; Du, H.W.; Liang, W.; Xu, C.X.; Lu, B.F. (2007). Effect of Mn on the Formation of Mg-based Spherical Icosahedral Quasicrystal Phase. *Journal of Alloys and Compounds*, Vol.427, pp. 244-250, ISSN 0925-8388.
[18] Mullins, W.W.; Swkerka, R.F. (1963). Morphological Stability of a Particle Growing by Diffusion or Heat Flow. *Journal of Applied Physics*, Vol.34, pp. 323-329, ISSN 0021-8979.
[19] Wang, Z.F.; Zhao W.M.; Li, H.P.; et al. (2010). Fabrication of Quaternary Mg-Zn-Y-Ce Quasicrystal Alloys and Their Strengthening Effects on AZ91 Magnesium Alloys. *China Foundry*, Vol.7, pp. 342-348, ISSN 1672-6421.
[20] Zhang, J.S.; Pei,L.X.; Du, H.W.; Liang, W.; Xu, C.X.; Lu, B.F. (2008). Effect of Mg-based Spherical Quasicrystals on Microstructure and Mechanical Properties of AZ91 Alloys. *Journal of Alloys and Compounds*, Vol.453, pp. 309-315, ISSN 0925-8388.
[21] Wang, Z.F.; Zhao W.M.; Qin, C.L.; et al. (2012). Direct Preparation of Nano-Quasicrystals via a Water-Cooled Wedge-Shaped Copper Mould. *Journal of Nanomaterials*, Vol.2012, Article ID 708240, ISSN 1687-4110.
[22] Wang, Z.F.; Zhao, W.M.; Qin, C.L.; Cui, Y. (2012). Fabrication and Corrosion Resistance of Mg-Zn-Y-based Nano-quasicrystals Alloys. *Materials Research*, Vol.15, pp. 51-56, ISSN 1516-1439.
[23] Wang, Z.F.; Zhao W.M.; Lin, X.P. ; et al. (2009). Effect of Mn and Cu on Spheroidized Process of Mg-Zn-Y-based Icosahedral Quasicrystal Phase. *Advanced Materials Research*, Vol.79-82, pp. 1403-1406, ISSN 1022-6680.
[24] Wang, Z.F.; Zhao W.M.; Li, H.P.; et al. (2011). Effects of Cooling Rate on Morphology, Microhardness and Volume Fraction of Mg-Zn-Y Quasicrystals. *Advanced Materials Research*, Vol.160-162, pp. 470-474, ISSN 1022-6680.
[25] Wang, Z.F.; Zhao W.M.; Li, H.P.; et al. (2011). Heat Treatment and Morphology Study of Mg-Zn-Y Quasicrystal Alloys. *Advanced Materials Research*, Vol.189-193, pp. 3824-3827, ISSN 1022-6680.

[26] Wang, Y.M.; Wang, Z.F.; Zhao W.M.; et al. (2011). Effects of Cooling Rate on Quasicrystal Microstructures of Mg-Zn-Y Alloys. *Advanced Materials Research*, Vol.160-162, pp. 901-905, ISSN 1022-6680.

[27] Zhao, W.M.; Wang, Z.F.; Li, H.P.; et al. (2011). Thermodynamics and Dynamics Study on Solidification Process of Mg_3Zn_6Y quasicrystals. *Advanced Materials Research*, Vol.214, pp. 128-132, ISSN 1022-6680.

[28] Zhang, J.; Qiu, K.; Wang, A.; et al. (2004). Effects of Pressure on the Solidification Microstructure of $Mg_{65}Cu_{25}Y_{10}$ Alloy. *Journal of Materials Science & Technology*, Vol.20, pp. 106-108, ISSN 1005-0302.

[29] Song, G.L. (2006). *Corrosion and Protection of Magnesium Alloys*, Chemistry Industry Press, ISBN 978-750-2585-65-5, Beijing, China. (In Chinese)

[30] Murty, B.S.; Hono, K. (2004). On the Criteria for the Formation of Nanoquasicrystalline Phase. *Applied Physics Letters*, Vol.84, pp. 1674-1676, ISSN 0003-6951.

[31] Sterzel, R.; Dahlmann, E.; Langsdorf, A.; Assmus, W. (2000). Preparation of Zn-Mg-Rare Earth Quasicrystals and Related Crystalline Phases. *Materials Science and Engineering A*, Vol.294-296, pp. 124-126, ISSN 0921-5093.

[32] Inoue, A. (2000). Stabilization of Metallic Supercooled Liquid and Bulk Amorphous Alloys. *Acta Materialia*, Vol.48, pp. 279-306, ISSN 1359-6454.

[33] Dong, C.; Chen, W.; Wang, Y.; et al. (2007). Formation of Quasicrystals and Metallic Glasses in Relation to Icosahedral Clusters. *Journal of Non-Crystalline Solids*, Vol.353, pp. 3405–3411, ISSN 0022-3093.

[34] Inoue, A.; Tsai, A.P.; Kimura, H.M.; Masumoto, T. (1988). Enthalpy Relaxation Behaviour of Al-Si-Cr Quasicrystalline and Amorphous Alloys upon Annealing. *Journal of Materials Science*, Vol.23, pp. 429–437, ISSN 0022-2461.

Section 2

Surface Treatments of Magnesium Alloys

Chapter 2

Investigation of the Structure and Properties of PVD and PACVD-Coated Magnesium Die Cast Alloys

Tomasz Tański

Additional information is available at the end of the chapter

http://dx.doi.org/10.5772/48165

1. Introduction

The development of existing techniques and technologies depends mainly on the used materials. The applied material determines the manufacturing method for the products. Criteria, which are chose by engineers and technologists for a proper material choice are among others: strength, hardness, elongation, density, corrosion resistance, ability for plastic deformation, or even recycling ability [1-9]. At present engineers have in their disposal modern equipment for investigation of the material structure, which allows it to perform a better and more efficient analysis of the mechanisms responsible for specific properties.

In recent years, there is visible an increasing interest on light metals, and especially materials with low density and relatively high strength properties [1-9]. The group of these materials include particular magnesium and its alloys. Mechanical properties of pure magnesium are relatively low and depends on its purity. In as cast state the tensile strength R_m is in the range of 80 - 120 MPa, yield strength Re = 20 MPa, elongation A = 4 - 6%, and hardness is equal 30 HB. Magnesium is recognised in the periodic table of elements in the group of earth alkali metals and is therefore not to found in nature in the basic form, but only in the form of chemical compounds. As a pure element magnesium has not found wide application possibilities, but as an alloy in combination with other elements such as aluminium, zinc, manganese, silicon, zirconium, thorium, lithium, calcium and rare earth metals, magnesium forms alloys with unique properties, which are used for production of diverse devices and machine- and construction elements. Magnesium alloys are characterised with the lowest density among all non-ferrous metals, as well with a favourable strength to density ratio, which means, that with a lower weight magnesium can hold similar static and dynamic loads compared to aluminium alloys, or even iron alloys. Moreover, these metals have a high vibration damping capacity and good casting properties, similar to these of aluminium cast alloys. Magnesium alloys are used for production of different types of car accessories such as steering wheels, pedals, wheels, seat frames, inlet manifolds, gear housings and other elements (Table 1) [1-9].

Stability of raw material cost and supply	the abundance of natural resources together with the cost effective primary processes contribute to a stable supply that can rapidly grow to meet future demand
Lower weight	magnesium is the lightest structural metal. It is lighter than aluminum by 35 percent and lighter than zinc by 73 percent. Its excellent strength-to-weight ratio significantly decreases the weight and cost of magnesium components
Durability	magnesium is a durable metal with excellent capacity for damping vibrations
Machinability	magnesium is the easiest metal to machine, which leads to rapid processing and the cost effective production of finished parts
Perfect for complex applications	in the die casting process magnesium can be formed into very complicated and thin-walled parts with a high degree of precision
Shielding against electromagnetic radiation	magnesium alloys, due to their excellent conductivity, shield against harmful electromagnetic radiation and are an ideal material to be used in electronic equipment, where electromagnetic radiation is an issue
Market growth	the increase in the use of magnesium applications is approximately 15-20 percent annually

Table 1. Magnesium alloys properties [1,2,8]

Increasingly, they find their application in the sports industry. They are also used for production of bicycle frames, ski and other sports equipment, etc. Beside the automotive and sport industry these alloys are used for construction elements, machinery and equipment, industrial automation, electronics, also in the military industry, and in the electrical branches (Table 1). Magnesium alloys are recognized as materials with high potential application possibilities, what is increasingly reflected in its increasing popularity and usage in production. Thanks to innovatory technologies, it is possible to perform castings from a few grams to several kilograms in weight. These alloys due to the number of its benefits will soon become an indispensable standard in the automotive, sports and aviation industry [1-9]. The growing trends in the production of magnesium alloys point at the increased necessity of their application in the world constructional industry, and the magnesium alloys will become one of the most frequent materials used in the following decades.

Optimization of chemical composition and technological conditions, production, casting and heat treatment of light metal alloys, due to the excellent set of mechanical properties and corrosion resistance based on the analysed precipitation- and phase transitions processes occurring in the investigated alloys during their cooling process is just one of the existing possibilities applicable for the enhancement of mechanical properties. Taking into account, that some properties are of interest only for the surface of the material, investigations were carried out concerning surface treatment of the magnesium alloys by applying of the physical vapour deposition processes. Due to insufficient stiffness of the substrate material a dual coating system was applied with a variable coating hardness, consisting of a soft ground - transition gradient layer - and a hard, wear-resistant outer layer. Thin hard PVD coatings on a soft surface, seams to be a preferred combination from the tribological point of view. Such coatings have found its special application for components working tribo-corrosive environment [10-16].

The aim of this research work is to determine the nature and properties of two-layer coating in a gradient like and hard wear resistant layer configuration, produced using the PVD and PACVD process on magnesium alloy substrate. Taking into account the optimisation of chemical composition and the development of optimal production conditions for achieving highest possible mechanical properties, corrosion resistance and functional properties compared to the existing standard surface layers.

2. Experimental procedure

The investigations have been carried out on test pieces of MCMgAl12Zn1 (Table 2, No. 1), MCMgAl9Zn (Table 2, No.2), MCMgAl6Zn (Table 2, No. 3), MCMgAl3Zn1 (Table 2, No. 4) magnesium alloys in as-cast and after heat treatment states. The chemical composition of the investigated materials is given in Table 2.

The mass concentration of main elements, %							
No.	Al	Zn	Mn	Si	Fe	Mg	Rest
1	12.1	0.62	0.17	0.047	0.013	86.96	0.0985
2	9.09	0.77	0.21	0.037	0.011	89.79	0.0915
3	5.92	0.49	0.15	0.037	0.007	93.33	0.0613
4	2.96	0.23	0.09	0.029	0.006	96.65	0.0361

Table 2. Chemical composition of investigated alloys, %

The coating deposition process of the coatings: Ti/TiCN/TiAlN and Cr/CrN/CrN was made in a device based on the cathodic arc evaporation method in an Ar, N_2 and C_2H_2 atmosphere moreover the DLC coating were deposited using acetylene (C_2H_2) as precursor and was produced by PACVD process (Table 3). Cathodes containing pure metals (Cr, Ti) and the TiAl (50:50 at. %) were used for deposition of the coatings. The diameter of the used cathodes was 65 mm. After pumping the chamber the base pressure was 5x10-3 Pa (Table 3).

The temperature was controlled by thermocouples. Then the substrates were cleaned by argon ion at the pressure 2 Pa for 20 min. To improve the adhesion of coatings, a transition Cr or Ti interlayer was deposited. The working pressure during the deposition process was 2-4 Pa depending of the coatings type. The distance between each of the cathodes and the deposited substrates was 120 mm.

The examinations of thin foils microstructure and phase identification were made on the JEOL 3010CX transmission electron microscope (TEM), at the accelerating voltage of 300 kV using selected area diffraction method (SAD) for phase investigations. The diffraction patterns from the TEM were solved using a special computer program "Eldyf" software supplied by the Institute of Material Science o the University of Silesia. TEM specimens were prepared by cutting thin plates from the material. The specimens were ground down to foils with a maximum thickness of 80 μm before 3 mm diameter discs were punched from the specimens. The disks were further thinned by ion milling method with the Precision Ion Polishing System (PIPS™), used the ion milling device model 691 supplied by Gatan until one or more holes appeared. The ion milling was done with argon ions, accelerated by a voltage of 15 kV, energy and angle are presented in Table 4.

Process parameters	Coating type		
	Ti/TiCN-gradient/TiAlN	Cr/CrN-gradient/CrN	Ti/DLC/DLC
Base pressure [Pa]	5×10^{-3}	5×10^{-3}	$1\text{x}10^{-3}$
Working pressure [Pa]	9.0◎10-1/1.1÷1.9/2.8	1.0/1.4÷2.3/2.2	2
Argon flow rate measurement [sccm]	80*	80*	80*
	10**	80**	-
	10***	20***	-
Nitrogen flow rate measurement [sccm]	0→225**	0→250**	-
	350***	250***	-
Acetylene flow rate measurement [sccm]	140→0**	-	230
Substrate bias voltage [V]	70*	60*	500
	70**	60**	
	70***	60***	
Target current [A]	60	60	-
Process temperature [ºC]	<150	<150	<180

Table 3. Deposition parameters of the investigated coatings
*during metallic layers deposition, **during gradient layers deposition, *** during ceramic layers deposition

Angle [°]	Energy [KeV]	Time [min]
6	3.8	180
3	3.2	15

Table 4. Ion milling parameters using for polishing

Microstructure investigation was performed using scanning electron microscope (SEM) ZEISS Supra 25 with a magnification between 10000 and 35000 times. For microstructure evaluation the Secondary Electrons (SE) detection was used, with the accelerating voltage of 5÷25 KV. For a complex metallographic analysis of the fractures of the investigated samples, the material with the coated layer was initially cut, and before braking cooled down in liquid nitrogen. Qualitative and quantitative chemical composition analysis in micro-areas of the investigated coatings was performed using the X-Ray microanalysis (EDS) by mind of the spectrometer EDS LINK ISIS supplied by Oxrord. This device is attached to the electron scanning microscope Zeiss Supra 35. The investigations were performed by an accelerating voltage of 20 kV.

The analysis of phase composition of the substrates and of the obtained coatings was carried out using the X-ray diffraction method (XRD) on the X-ray apparatus X'Pert of the Panalytical Company using the filtered radiation of a cobalt lamp. For the reason of put on diffraction pattern coming from the substrate material and coating ones as well as their intensity, and also convergence of the Miller indexes for different coating types to achieve a more accurate information from the surface layer and decrease of the substrate influence on the diffraction pattern in the future investigations a diffraction technique was applied with a constant angle of incidence of the primary X-Ray beam using parallel beam collimator placed before the proportional detector. Diffraction pattern of gradient- and multilayer coatings were achieved by different angle of incidence of the primary beam. The effective depth of measurement was chosen experimentally on the basis of preliminary investigations of different angles of incidence.

The specimens were tested on Raman spectroscope from Renishaw, type inVia Raman Microscope, coupled with light microscope; for observation has been used laser with wavelength 514nm and full power equal to 30mW; after precise alignment of the laser spot the data was obtained in single spectral acquisition with use of 20x long working distance plan objective lens.

The cross-sectional atomic composition of the samples (coating and substrate) was obtained by using a glow discharge optical spectrometer, GDOS-750 QDP from Leco Instruments. The following operation conditions of the spectrometer Grimm lamp were fixed during the tests:

- lamp inner diameter – 4 mm;
- lamp supply voltage – 700 V;
- lamp current – 20 mA;
- working pressure – 100 Pa.

Wear resistance investigations were performed using the ball-on-disk method in dry friction conditions in horizontal settlement of the rotation axis of the disk. As the counterpart there was used a tungsten carbide ball with a diameter of 3 mm. The tests were performed at room temperature by a defined time using the following test conditions:

- load, Fn-5N,
- rotation of the disk 200 turns/min,
- wear radius of 2.5 mm,
- shift rate of -0, 05 m/s.

The measurement of roughness of the surface of the obtained coatings was carried out using the profilographometer Diavite Compact of Asmeo Ag Company. The measurement length of Lc=0.8 mm and measurement accuracy of ±0.02 μm. The parameter Ra acc. the Standard PN-EN ISO 4287:1999 was assumed as the quantity describing the roughness. On each investigated sample there was performed 6 measurements and we determined the average.

3. Results and discussions

Results of diffraction method investigations achieved by the high resolution transmission electron microscope allow to identify the TiAlN, CrN, graphit phase occurred in the surface layer (Fig. 1-12). For all investigated alloys a nanocrystalline microstructure of the surface layer in investigated area was detected. On figure 1, 2, there are showed the microstructures of the layer TiAlN phase, using the dark field technique the size of the subgrains or crystallites can be determined, as ca. 15 nm in diameter. For phase determination of the structure of the surface layer diffraction pattern analysis of the investigated areas has allow it to identify the (Ti,Al)N phase as a cubic phase of the 225-Fm3m space group with the d-spacing of a=b=c=0,424173 nm. The CrN phase was determined as a cubic phase of the 225-Fm3m space group with the d-spacing of a=b=c=0.414 nm. Investigations performed using particularly the dark field technique on the transmission electron microscope have confirmed that the size of the CrN crystallites, in the majority of the cases does not exceed the limit of ~ 20 nm (Fig. 5,6). Also a globular bulk shaped morphology and homogeneity of these crystallites was found, as well a low statistical dispersion in the range between 10 to 20 nm.

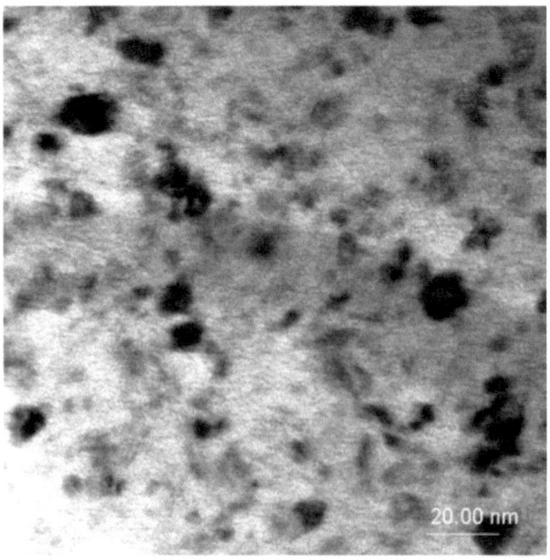

Figure 1. Structure of the thin foil from TiAlN surface layer (Ti/TiCN/TiAlN coating), bright field, TEM

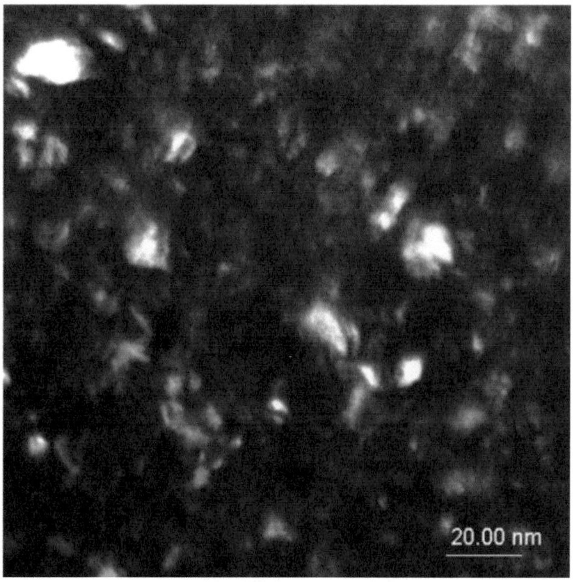

Figure 2. Structure of the thin foil from TiAlN surface layer (Ti/TiCN/TiAlN coating), dark field, TEM

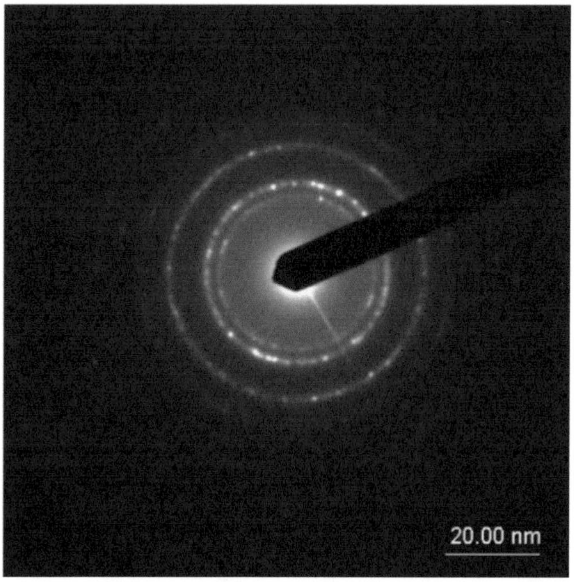

Figure 3. Diffraction pattern of the thin foil from TiAlN surface layer (Ti/TiCN/TiAlN coating) presented on Fig. 1

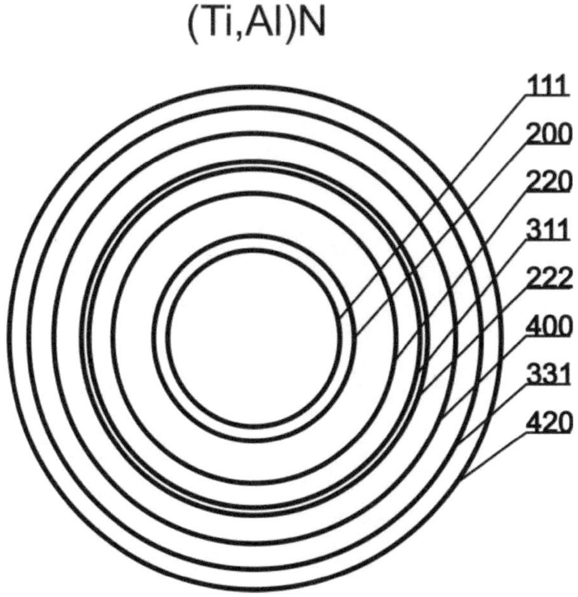

Figure 4. Solution of the diffraction pattern presented on Fig. 3 for the (Ti, Al)N phase

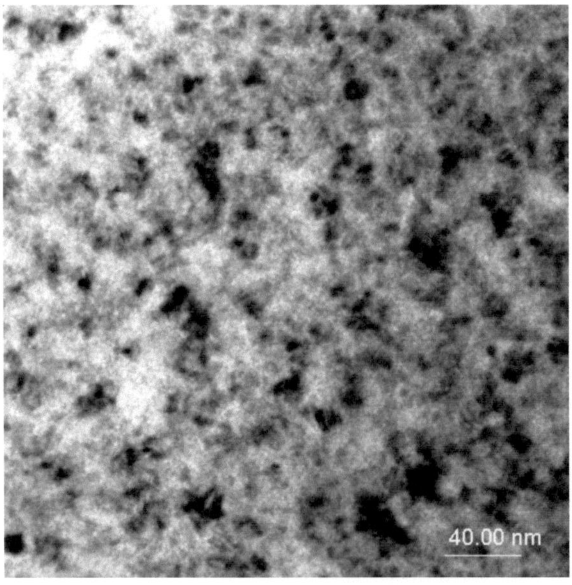

Figure 5. Structure of the thin foil from CrN surface layer (Cr/CrN/CrN coating), bright field, TEM

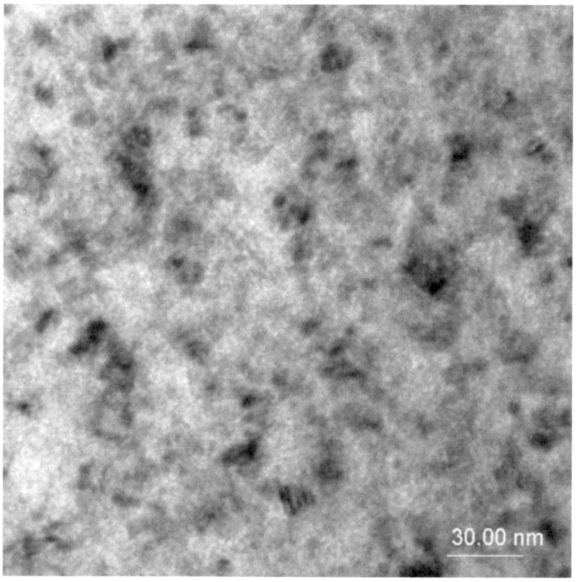

Figure 6. Structure of the thin foil from CrN surface layer (Cr/CrN/CrN coating), bright field, TEM

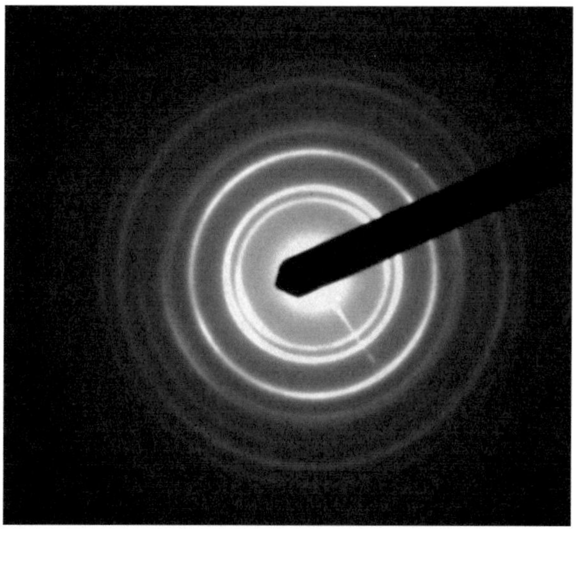

Figure 7. Diffraction pattern of the thin foil from CrN surface layer (Cr/CrN/CrN coating) presented on Fig. 5

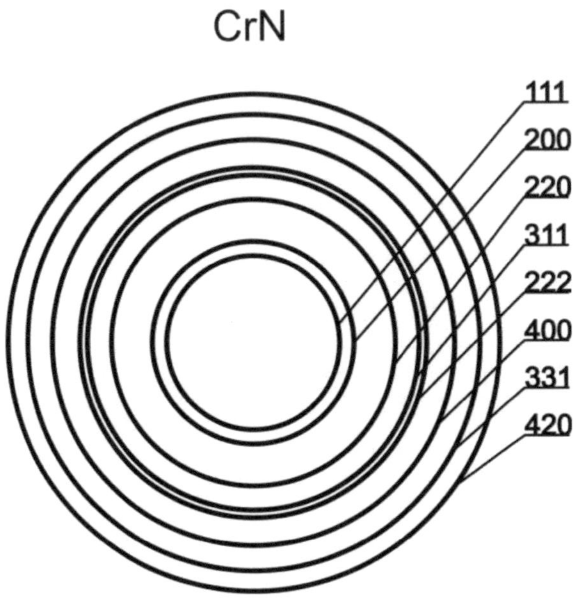

Figure 8. Solution of the diffraction pattern presented on Fig. 7 for the CrN phase

The graphite phase was determined as a hexagonal phase of the 186-P63mc space group with the lattice parameters of a=b=0.2, c=0.679 nm (Fig. 9,10). Investigations performed using particularly the dark field technique on the transmission electron microscope have confirmed, that the size of the graphite crystallites, which the Ti/DLC/DLC coating is constant, is in the range up to 30 nm and with irregular shape.

Investigations of the fractures of the magnesium alloys coated with the Ti/TiCN/TiAlN, Cr/CrN/CrN, Ti/DLC/DLC layers show an occurrence of sharp transition zone between the substrate and coating (Fig. 13-18). It was found out, as a result of the microstructure investigations on scanning electron microscope, that there are no pores or cracks in the produced coating and no defects and failures occurring spontaneously in this single layer are of significant importance for the properties of the whole layer (Fig. 13-18). The thickness of the Ti/TiCN/TiAlN layer is in the range up to 3.3 μm, Cr/CrN/CrN layer is in the range up to 1.9 μm, and Ti/DLC/DLC layer is in the range up to 2,5 microns.

In the case of the Ti/TiCN/TiAlN layer it was also found that the examined layers were not uniform and consisted of three sub-layers, with a clearly visible transition zone between the gradient layer and the wear resistant coating achieved using separate metals evaporation sources, where the upper one had a thickness of ca. 0.6 μm (Fig. 13, 14). Coating thickness was measured using a scanning electron microscope. Fracture morphology of the investigated coatings is characterised by a lack of columnar structure (Fig. 13-18). On the basis of the performed observations on scanning electron microscope the coating of the Ti/TiCN/TiAlN type show an increasing non-homogeneity compared to the Cr/CrN/CrN and Ti/DLC/DLC coatings what is connected with the presence of numerous droplet-shaped microparticles (their number depends on the type of the target) and should that fore significantly influence mechanical properties and resistance of the investigated surfaces (Fig. 13-20). The droplets observed in SEM are noticeably different in terms of size and shape (regular and irregular shape, slightly flat). There were also some hollows formed probably when the solidified droplets break off after the used process has been completed (Fig. 19,20).

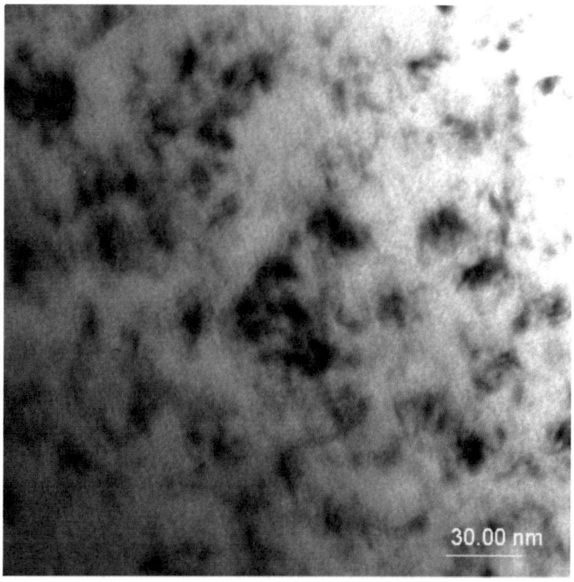

Figure 9. Structure of the thin foil from DLC surface layer (Ti/DLC/DLC coating), bright field, TEM

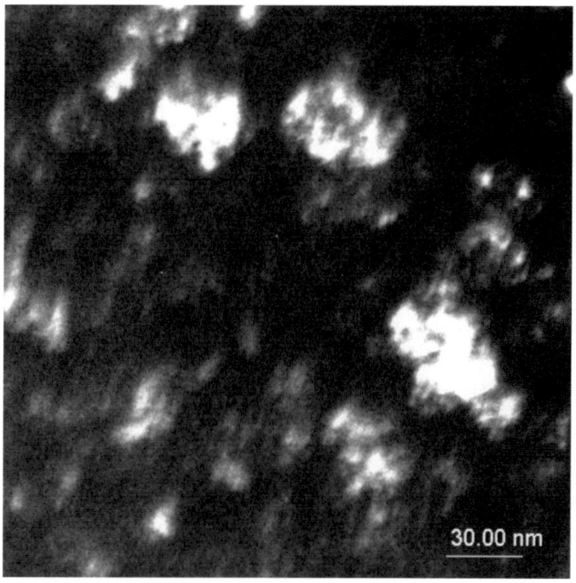

Figure 10. Structure of the thin foil from DLC surface layer (Ti/DLC/DLC coating), dark field, TEM

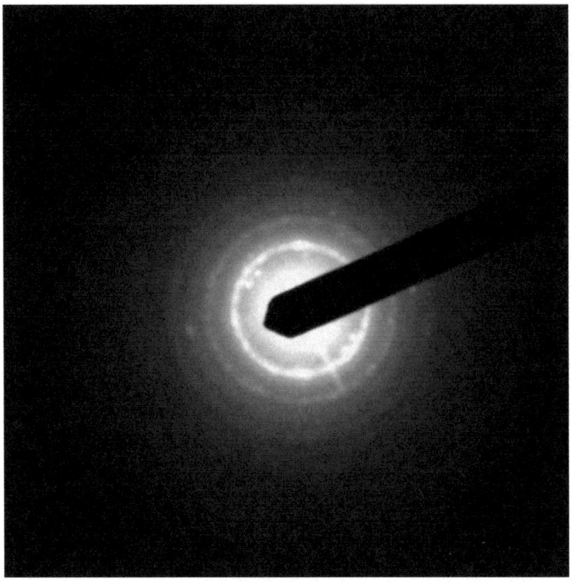

Figure 11. Diffraction pattern of the thin foil from DLC surface layer (Ti/DLC/DLC coating) presented on Fig. 9

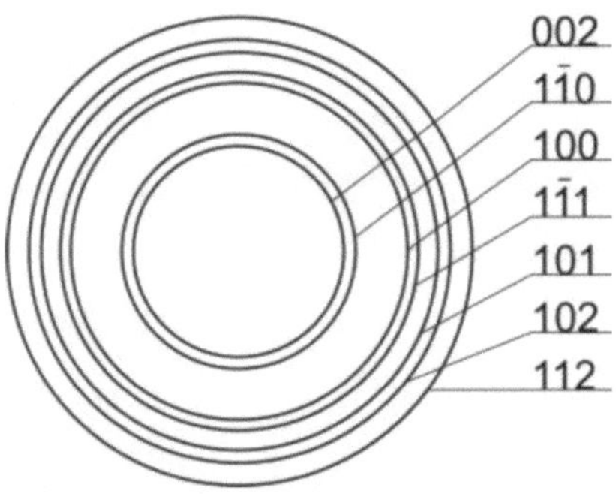

Figure 12. Solution of the diffraction pattern presented on Fig. 11 for the graphite phase

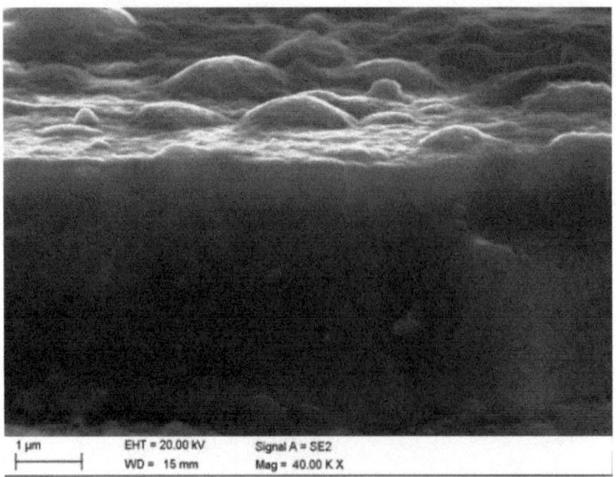

Figure 13. Cross-section SEM images of the Ti/TiCN/TiAlN coating deposited onto the AZ91 substrate

Figure 14. Cross-section SEM images of the Ti/TiCN/TiAlN coating deposited onto the AZ91 substrate

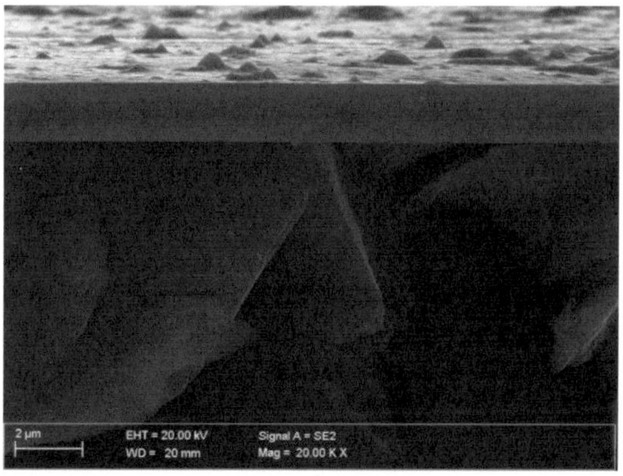

Figure 15. Cross-section SEM images of the Cr/CrN/CrN coating deposited onto the AZ121 substrate

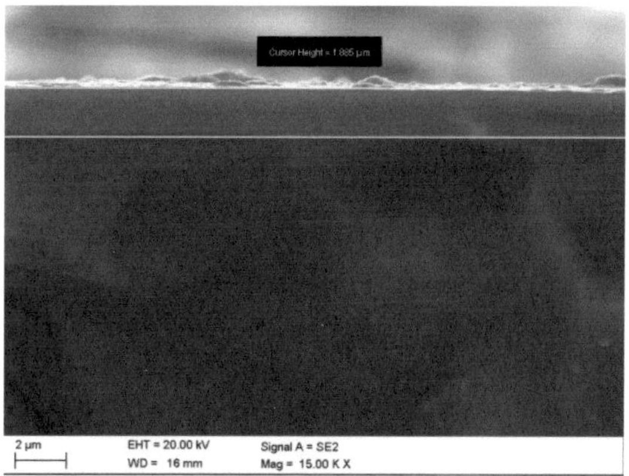

Figure 16. Cross-section SEM images of the Cr/CrN/CrN coating deposited onto the AZ61 substrate

As a result of the performed X-ray microanalysis using the qualitative energy spectrometer EDS it was confirmed the presence of major alloying elements Mg, Al, Zn, N, Cr, as compounds of the investigated alloys as well of the coatings (in this case Cr/CrN/CrN coating) (Fig. 21). Moreover qualitative analysis of the chemical elements distribution performed on the cross-section of the investigated sample clearly confirms increase of the concentration of the elements at grain boundaries of the produced coatings (Fig. 21).

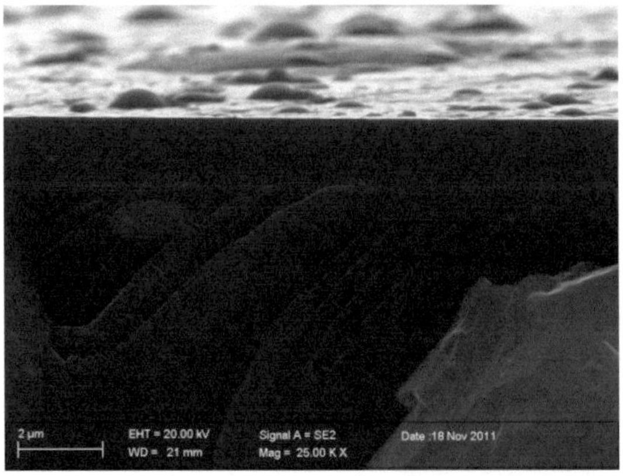

Figure 17. Cross-section SEM images of the Ti/DLC/DLC coating deposited onto the AZ121 substrate

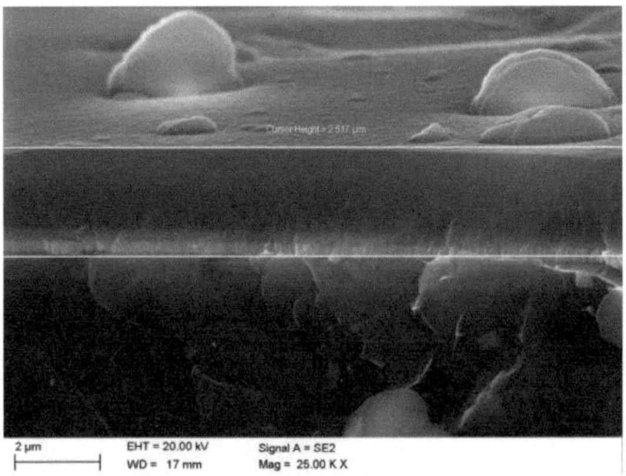

Figure 18. Cross-section SEM images of the Ti/DLC/DLC coating deposited onto the AZ61 substrate

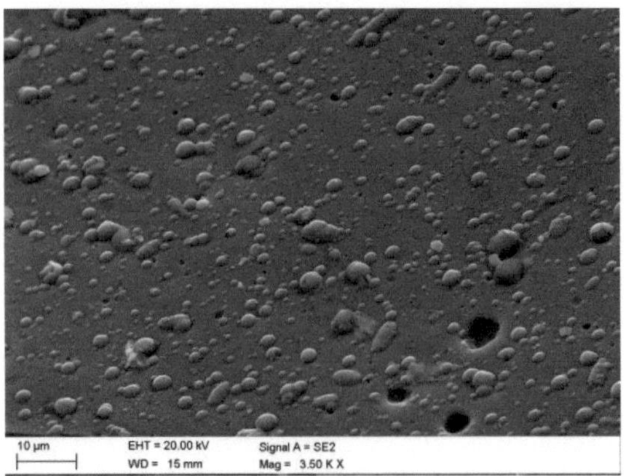

Figure 19. Surface topography of the Ti/TiCN/TiAlN coating deposited onto AZ61 substrate

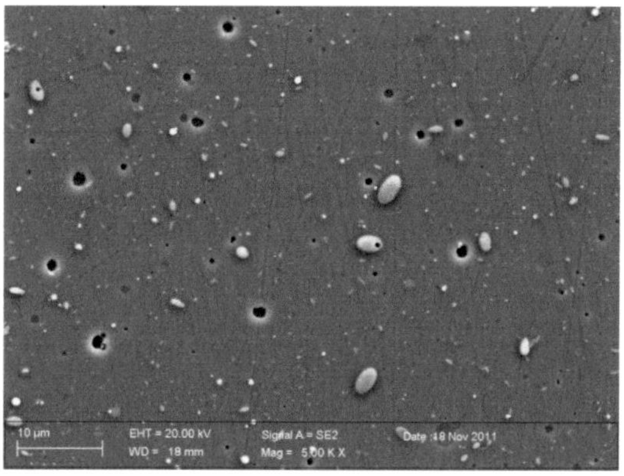

Figure 20. Surface topography of the Cr/CrN/CrN coating deposited onto AZ91 substrate

Based on the results obtained by the quantitative X-ray microanalysis using the energy dispersed X-ray EDS spectrometer it was confirmed the presence of Mg, Al, Zn, Ti, C as major alloying elements of the cast magnesium alloys as well the obtained coatings (Fig. 22, Table 5). Due to the fact, that the EDS analysis in case of measurement of so called light element concentrations, for which the energy <1 keV (C) has a relatively large measurement error because of strong absorption. For this reason the described values should be seen as estimated values only. But the measurement error in case of mass concentration measurement - in the range from 5 to 20 % - is about 2%.

Chemical element	The mass and atomic concentration of main elements, %	
	mass	atomic
	Analysis 1 (point 1)	
C	92.85	96.92
Mg	04.33	02.24
Al	00.52	00.24
Ti	02.29	00.60
Matrix	Correction	ZAF
	Analysis 2 (point 2)	
C	76.59	89.31
Zn	00.84	00.18
Mg	12.56	07.23
Al	01.55	00.81
Ti	08.46	02.47
Matrix	Correction	ZAF
	Analysis 2 (point 3)	
Zn	05.67	02.25
Mg	67.38	71.85
Al	26.95	25.90
Matrix	Correction	ZAF

Table 5. The results of quantitative chemical analysis from third 1, 2, 3 areas of coating Ti/DLC/DLC deposited onto substrate from AZ91 alloy marked in Fig. 22

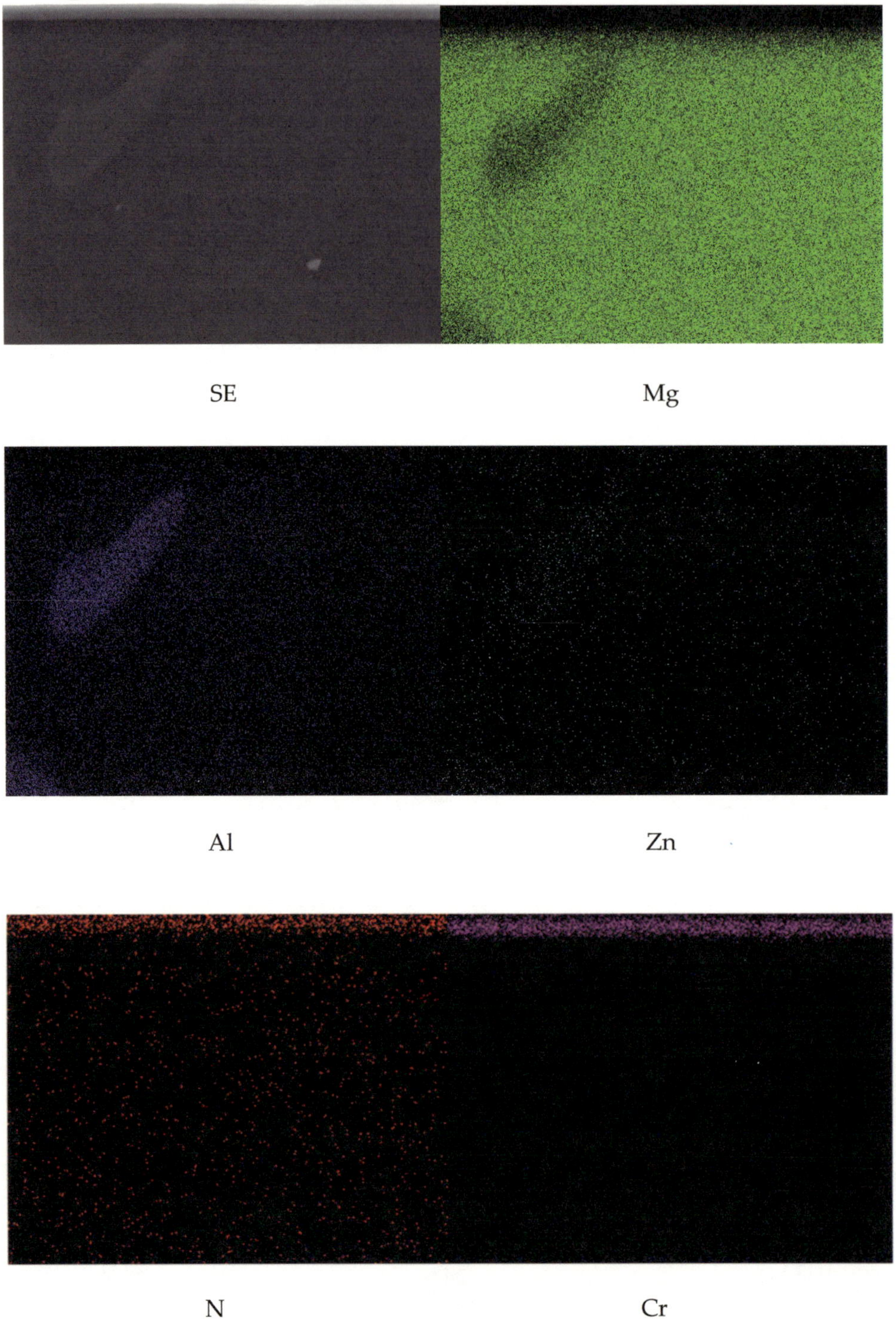

Figure 21. The area analysis of chemical elements of the Cr/CrN/CrN coating and the magnesium (AZ61) substrate: image of the secondary electrons (A) and maps of elements' distribution

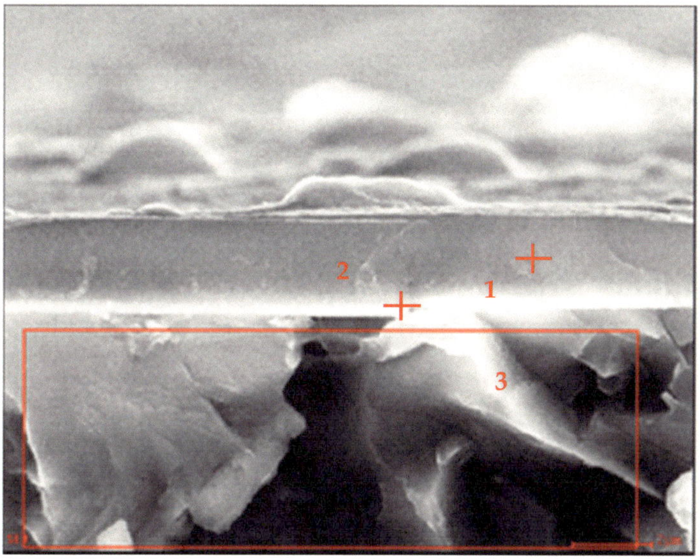

Figure 22. Cross-section SEM images of the Ti/DLC/DLC coating deposited onto the AZ91 substrate

Changes of coating component concentration and substrate material made in GDOS were presented in Figs. 23, 24. The tests carried out with the use of GDOS indicate the occurrence of a transition zone between the substrate material and the coating, which results in the improved adhesion between the coatings and the substrate. In the transition zone between the coatings and the substrate, the concentration of the elements of the substrate increases with simultaneous rapid decrease in concentration of elements contained in the coatings. The existence of the transition zone should be connected with high-energy ion action that caused mixing of the elements in the interface zone.

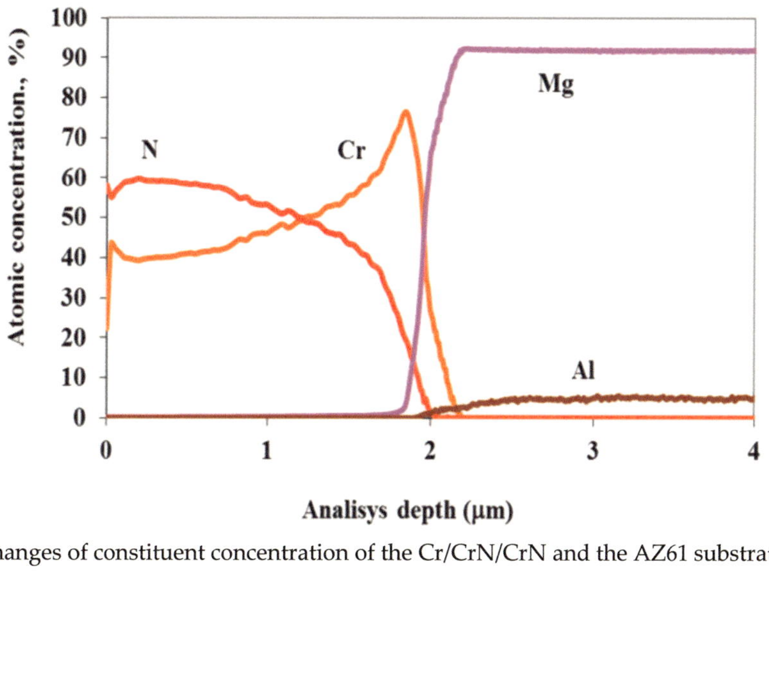

Figure 23. Changes of constituent concentration of the Cr/CrN/CrN and the AZ61 substrate materials

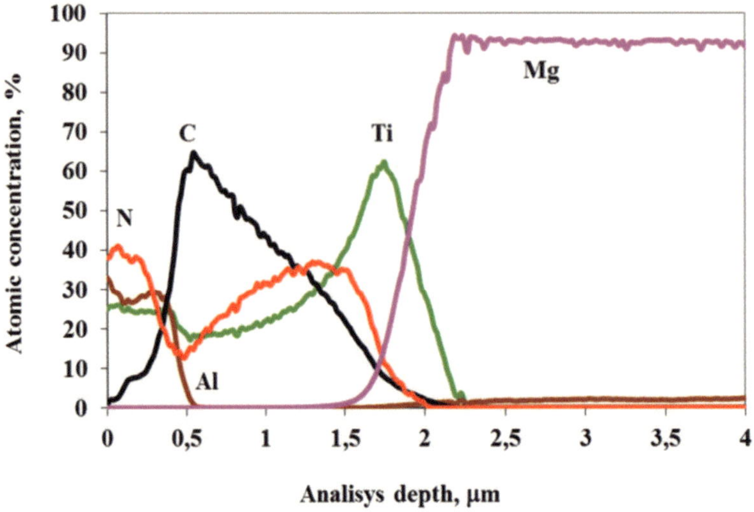

Figure 24. Changes of constituent concentration of the Ti/TiCN/TiAlN and the AZ61 substrate materials

On Figs. 25 a and b there are presented the X-ray diffractions of the investigated magnesium alloys in state after heat treatment. Using the of qualitative X-ray phase analysis methods it was confirmed, that in the investigated materials occurs the γ ($Mg_{17}Al_{12}$) phase as well the α-Mg phase which is the alloy matrix. A too small volume fraction of other phases present in the material does not allow it to perform an unambiguous identification of the obtained X-ray diffractions. Because of the overlapping reflections of the substrate and the coating material, as well the relatively small thickness of each layer, there were difficulties with identification of the phases. It was also confirmed the presence of reflexes coming from the phases present in the substrate, e.g. α and γ (Fig. 25 c,d). Very small volume fraction of other phases present in the substrate material does not allow it to perform an unambiguous identification of the recorded X-ray spectrum. The presence of substrate reflexes was confirmed on every achieved X-Ray diffraction collected from the coating, due to the thickness of the obtained coatings <3.5 µm, smaller than the X-ray penetration depth. Using the technique of fixed incidence angle (GIXRD method) there are collected only reflexes from the thin surface layers (Fig. 25 e,f).

The morphology of the deposited films, particulary DLC coating was characterized also by Raman spectroscopy. This spectroscopy method was used to determine the microstructure and chemical composition of the deposited DLC films. The shape of the achieving Raman spectrum is characteristic for carbon materials with a low level of structure order. The obtained spectrum can be presented in form of two Gaussian curves, respectively for the Raman shift values equal ca 1500 cm -1 (D band) and 1300 cm -1 (G band). The ratio of their height can be presented as a ordering level of the carbon structure of the material analyzed. The analyzed layer is composed of amorphous carbon - or more precisely- composed of poorly structured carbon material, including small crystallites.

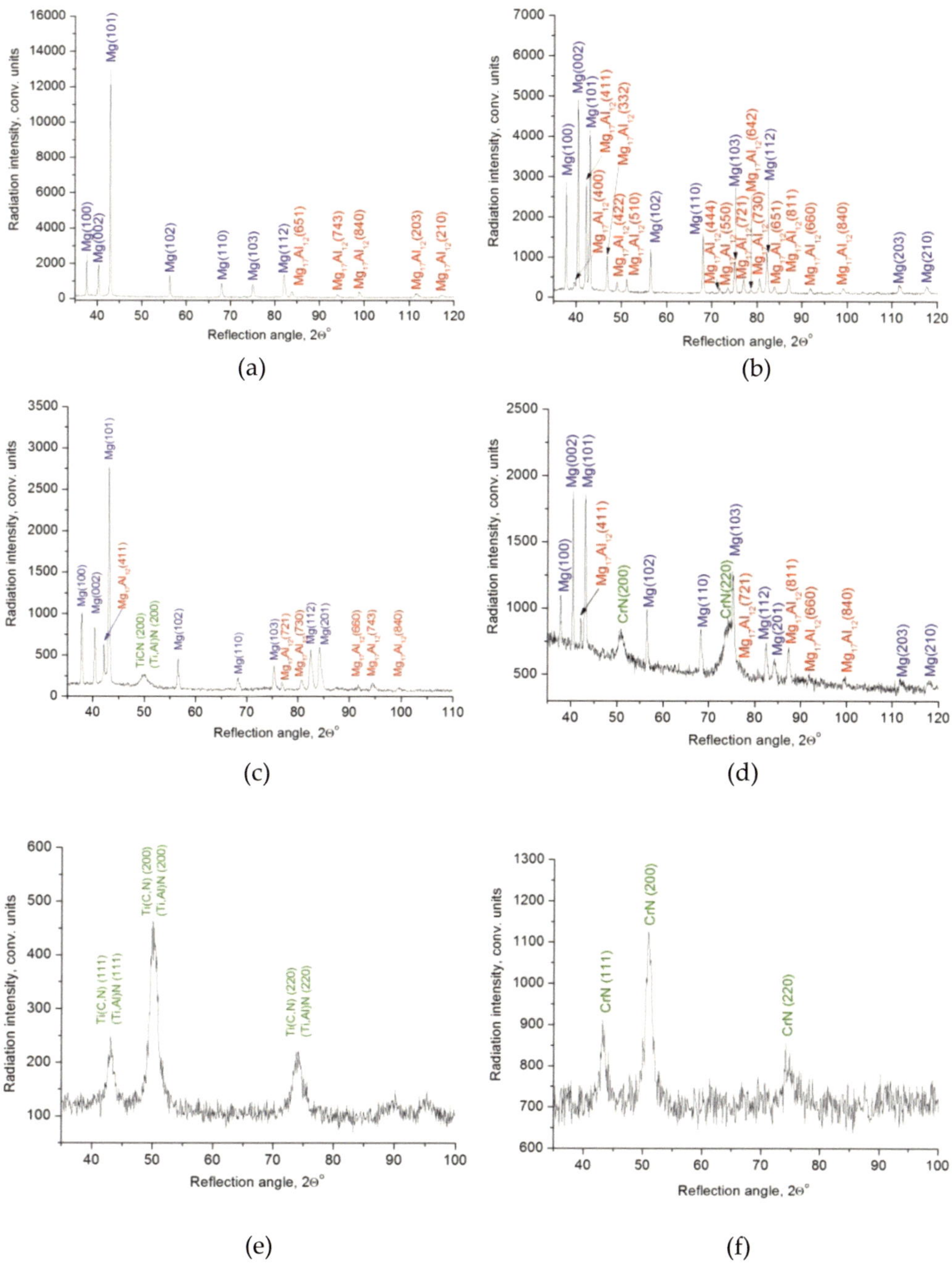

Figure 25. X-ray diffraction pattern of: a) AZ31 substrate, b) AZ121 substrate, c) Ti/TiCN/TiAlN coating deposited on the AZ91 magnesium alloys, d) Cr/CrN/CrN coating deposited on the AZ91 magnesium alloys obtained by Bragg-Brentano method, e) Ti/TiCN/TiAlN coating deposited on the AZ61 magnesium cast alloy obtained by GIXRD method (α=4º), f) Cr/CrN/CrN coating deposited on the AZ61 magnesium cast alloy obtained by GIXRD method (α=4º)

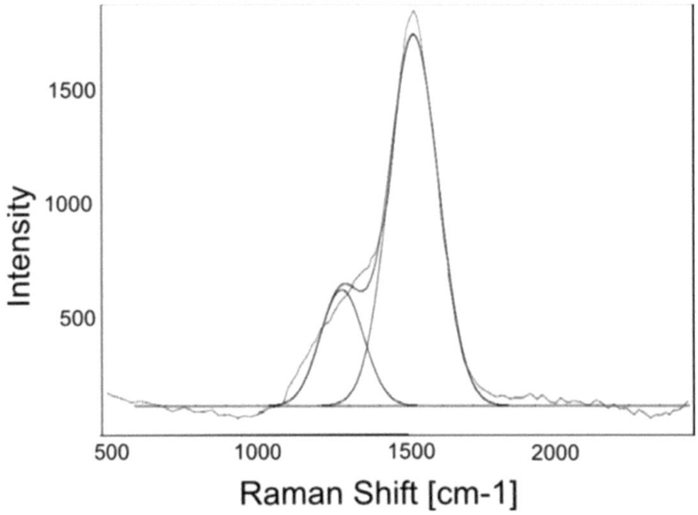

Figure 26. Raman spectra of the DLC films deposited on AZ91 magnesium alloys.

The highest value of surface roughness equal 0.3 mm was measured for the coating of the Ti/TiCN/TiAlN type which is likely caused by the occurrence of numerous microparticles in the shape of droplets in the structure (Table 6, Figure 13, 14, 19). The observed high homogeneity of the Cr/CrN/CrN surface coating is characterized by a smaller amount of crystallised droplets of liquid metal (Table 6, Figure 20), what responses to a smaller surface roughness within the range from 0.12 to 0.15 mm. The Ti/DLC/DLC coating has a surface roughness of ~ 0.25 mm. The performed investigations of the surface of the cast magnesium alloy, with coated layers confirm a lack of significant effect of the substrate type on the surface roughness (Table 6).

To determine the tribological properties of the investigated coating deposited on the magnesium alloys substrate, an abrasion test under dry slide friction conditions was carried out by the ball-on-disk method. Table 6 and Fig. 27 presents the friction coefficient and sliding distance results for each type of the investigated substrate. Under technically dry friction conditions, after the wearing-in period, the friction coefficient recorded for the associations tested is stabilized in the range 0.08-0.38 μm depending on the used substrate and coatings. All friction coefficient diagrams which were collected depending on the rotation rate or friction path length have similar characteristics and can be divided into two parts (Fig. 27). In the first part, there occurs a sharp increase of the friction coefficient together with increasing friction path length. It was assumed, that this is a transient state of the friction process. The second part of the graph has already a stable state. Rapid changes of the friction coefficient value are caused by the occurrence of pollutants in form of sample counterface spalling products (balls are made from WC), which disturb the measurement of the friction coefficient. Comparing the friction coefficient results with the friction path length, it was found that the best wear resistance is characteristic for materials coated with DLC carbon. According to the applied load of 5 N, the average friction coefficients for the DLC coatings with the sliding rate of 0.05 m / s is in the range of 0.08-0.15 mm, which is ten times lower compared to the friction coefficient values of other examined coatings. However, the results of the friction path length for the DLC coatings were at a level exceeding even 70 times the results of the friction path length achieved for the Cr/CrN/CrN coatings. This is characteristic for DLC coatings, because they are composed of poorly ordered graphite, which is probably formed by a friction-assisted phase transformation of the surface layer of the DLC matrix and acts as a lubricant at the surface [17]. Accordingly, the high hardness of DLC together with this transfer layer is responsible for the low friction coefficient of the DLC film in comparison with magnesium alloys coated other investigated coatings. At high sliding speed, the transfer layer is more easily formed due to the accumulation of heat, resulting in a lower friction coefficient.

		AZ121	
Coatings	**Roughness [μm]**	**Friction coefficient, [μm]**	**Sliding disance, [m]**
Cr/CrN/CrN	0.13	0.24-0.27	8
Ti/TiCN/TiAlN	0.28	0.19-0.22	59.4
Ti/DLC/DLC	0.24	0.09-0.16	550
		AZ91	
Coatings	**Roughness [μm]**	**Friction coefficient, [μm]**	**Sliding disance, [m]**
Cr/CrN/CrN	0.12	0.25-0.38	7.8
Ti/TiCN/TiAlN	0.27	0.18-0.25	57.6
Ti/DLC/DLC	0.25	0.1-0.17	540
		AZ61	
Coatings	**Roughness [μm]**	**Friction coefficient, [μm]**	**Sliding disance, [m]**
Cr/CrN/CrN	0.15	0.22-0.28	22
Ti/TiCN/TiAlN	0.30	0.17-0.22	77.7
Ti/DLC/DLC	0.25	0.09-0.19	630
		AZ31	
Coatings	**Roughness [μm]**	**Friction coefficient, [μm]**	**Sliding disance, [m]**
Cr/CrN/CrN	0.12	0.2-0.29	13
Ti/TiCN/TiAlN	0.28	0.15-0.22	66
Ti/DLC/DLC	0.26	0.08-0.15	605

Table 6. The characteristics of the tested coatings

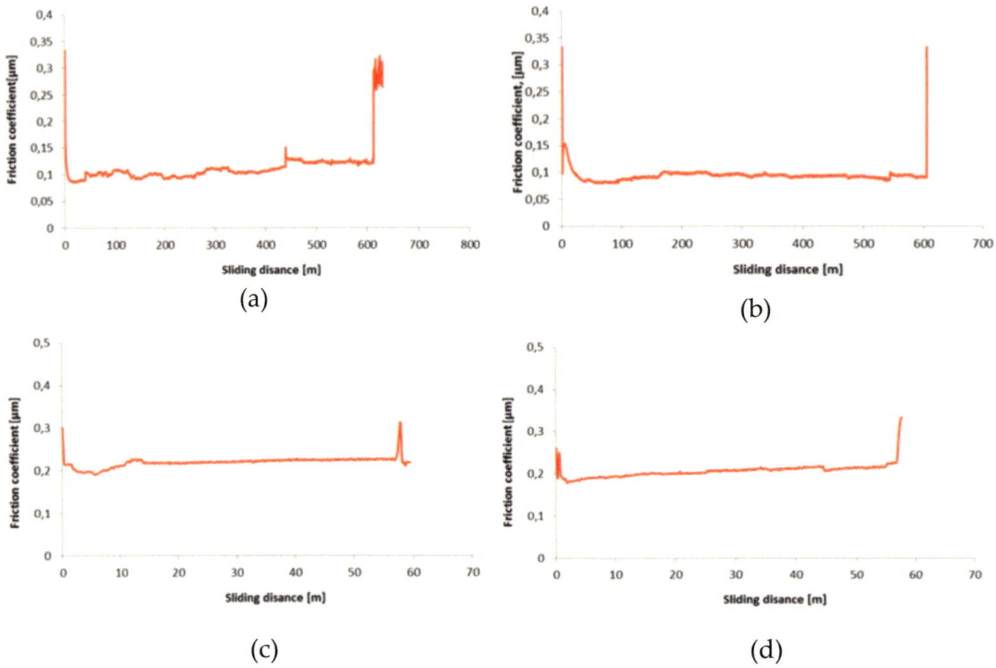

Figure 27. Dependence of friction coefficient on sliding distance during the wear test for:
a) Ti/DLC/DLC coating deposited on the AZ61, b) Ti/DLC/DLC coating deposited on the AZ31,
c) Ti/TiCN/TiAlN coating deposited on the AZ121, d) Ti/TiCN/TiAlN coating deposited on the AZ91

4. Summary

Due to the character of the investigated material (magnesium alloys) and its relatively low melting point, the whole technological PVD and PACVD processes were performed at temperatures up to 150 °C for Cr/CrN/CrN coatings and Ti/TiCN/TiAlN coatings, and up to 180° C for Ti/DLC/DLC coatings. Results of diffraction method investigations achieved by the high resolution transmission electron microscope allow to identify the TiAlN, CrN, graphit phase occurred in the surface layer. It was found out, as a result of the microstructure investigations on scanning electron microscope, that there are no pores or cracks in the produced coating and no defects and failures occurring spontaneously in this single layer are of significant importance for the properties of the whole layer. Coating thickness was measured using a scanning electron microscope. The thickness of the Ti/TiCN/TiAlN layer is in the range up to 3.3 μm, Cr/CrN/CrN layer is in the range up to 1.9 μm, and Ti/DLC/DLC layer is in the range up to 2.5 microns. The tests carried out with the use of GDOS indicate the occurrence of a transition zone between the substrate material and the coating, which results in the improved adhesion between the coatings and the substrate. Using the technique of fixed incidence angle (GIXRD method) there are collected only reflexes from the thin surface layers. The highest wear resistance was obtained for the of Ti/DLC/DLC coating.

Author details

Tomasz Tański
Faculty of Mechanical Engineering, Silesian University of Technology, Gliwice, Poland

Acknowledgement

Research was financed partially within the framework of the Polish State Committee for Scientific Research Project No. 4688/T02/2009/37 headed by Dr Tomasz Tański

5. References

[1] Horst EF, Mordike BL (2006) Magnesium Technology. Metallurgy, Design Data, Application, Springer-Verlag, Berlin Heidelberg, 2006. 707 p.
[2] Tański T, Dobrzański L A, Labisz K (2010) Investigations of microstructure and dislocations of cast magnesium alloys. Journal of Achievements in Materials and Manufacturing Engineering. 42/1-2: 94-101.
[3] Easton M, Beer A, Barnett M, Davies C, Dunlop G, Durandet Y, Blacket S, Hilditch T, Beggs P (2008) Magnesium Alloy Applications in Automotive Structures. Journal Minerals, Metals and Materials Society. 60:57-62.
[4] Mehta DS, Masood SH, Song WQ (2004) Investigation of wear properties of magnesium and aluminium alloys for automotive applications. Journal of Materials Processing Technology. 155-156:1526–1531.
[5] Dobrzański LA, Tański T, Čížek L (2006) Influence of Al addition on microstructure of die casting magnesium alloys. Journal of Achievements in Materials and Manufacturing Engineering 19: 49-55.
[6] Dobrzański LA, Tański T, Čížek L (2007) Heat treatment impact on the structure of die-cast magnesium alloys. Journal of Achievements in Materials and Manufacturing Engineering. 20:431-434.
[7] Tański T, Dobrzański LA, Čížek L (2007) Influence of heat treatment on structure and properties of the cast magnesium alloys. Journal of Advanced Materials Research. 15-17: 491-496.
[8] Dobrzański LA, Tański T (2009) Influence of aluminium content on behaviour of magnesium cast alloys in bentonite sand mould. Solid State Phenomena. 147-149: 764-769.
[9] Dobrzański L A, Tański T, Malara Sz, Król M, Domagała-Dubiel J (2011) Contemporary forming methods of the structure and properties of cast magnesium alloys. In Czerwinski F, editor. Magnesium Alloys - Design, Processing and Properties. Rijeka: InTech. pp. 321-350.
[10] Tański T, Labisz K (2012) Electron microscope investigation of PVD coated aluminium alloy surface layer. Solid State Phenomena 186: 192-197.
[11] Tański T., Lukaszkowicz K (2011) Structure and mechanical properties of hybrid-layers coated applying the PVD method onto magnesium and aluminium alloys substrate. Materials Engineering. 4:772-775.

[12] Lukaszkowicz K, Czyżniewski A, Kwaśny W, Pancielejko M (2011) Structure and mechanical properties of PVD coatings deposited onto the X40CrMoV5-1 hot work tool steel substrate. Vacuum. 2011 in print.

[13] Hollstein F, Wiedemann R, Scholz J (2003) Characteristics of PVD-coatings on AZ31HP magnesium alloys. Surface and Coatings Technology. 162:261–268.

[14] Veprek S, Veprek-Heijman MJG (2008) Industrial applications of superhard nanocomposite coatings. Surface & Coatings Technology. 202:5063-5073.

[15] Alvarez J, Melo D, Salas O, Reichelt R, Oseguera J, Lopez V (2009) Role of Al oxide PVD coatings in the protection against metal dusting. Surface & Coatings Technology. 204:779–83.

[16] Dobrzański LA, Staszuk M (2010) PVD and CVD gradient coatings on sintered carbides and sialon tool ceramics. Journal of Achievements in Materials and Manufacturing Engineering. 43/2: 552-576.

[17] Zou YS, Wu YF, Yang H, Cang K, Song GH, Li ZX, Zhou K (2011) Applied Surface Science. 258:1624-1629.

Chapter 3

Technology Foresight Results Concerning Laser Surface Treatment of Casting Magnesium Alloys

Anna Dobrzańska-Danikiewicz, Tomasz Tański,
Szymon Malara and Justyna Domagała-Dubiel

Additional information is available at the end of the chapter

http://dx.doi.org/10.5772/39208

1. Introduction

In accordance with the concept presented, innovations, understood as valuable, innovative ideas, should be the way to achieve economic growth and to solve the contemporary problems of the climate change, increased consumption and depletion of conventional energy sources, food security, healthcare, and the advancing ageing of societies. To tackle this challenge, the European Commission has formulated the Europe 2020 strategy [1] and has set up the Innovation Union [2]. It is estimated that the level of R&D and innovation investments until 2020 is to reach aggregately 3% of the EU's GDP from public and private funds. In order to achieve satisfactory economic and social effects, the stream of investments should be channelled into those fields of science and industries bringing the highest added value, with special consideration given to the role of small and medium sized enterprises. The aim of foresight research conducted broadly in Europe and Poland, also in the field of material engineering, is a quest for innovative areas deserving financial support [3-7]. Technology foresight serving to identify the priority, innovative technologies and the directions of their strategic development was pursued for materials surface engineering, as well [8]. One of 14 thematic areas analysed under such foresight research are laser technologies in surface engineering. Laser remelting and alloying / cladding is one of the critical technologies having the best development prospects and/or being of key significance for the industry selected for the detailed research carried out with the Delphi method.

Magnesium alloying with aluminium, manganese, metals of rare earths, thorium, zinc and zirconium enhances the strength in relation to the mass index, hence making them an important material where a decreased mass is important and where the forces of inertia must be reduced [9-11]. The advantages of the laser surface treatment processes, i.e.: short process time, flexibility, as well as operational precision, offer the upper hand of this method against the other methods employed in surface engineering. The primary aim of the laser remelting of material surface layers is to modify the structure and the associated properties [9-15]. Heightened resistance to, notably, wear and thermal fatigue is gained by creating a chemically homogenous, fine crystalline surface layer without changing the chemical composition of the material. Even more advantageous effects, such as improved functional properties, are feasible through alloying the material surface layer with the particles of the hard phases of carbides, oxides and nitrides. A need for magnesium alloys stems mainly from the development of the automotive and aviation industry. Rapid growth in the use of magnesium and magnesium alloys nearly in all the fields of the contemporary industry has been seen in the recent decades due to the numerous properties of this metal enabling to use it as a structural component as well as an additive to other chemical metal alloys. It is 35% lighter than aluminium (2.7g/cm3) and over four times lighter than steel (7.86g/cm3) [9-15,18]. Magnesium alloys, besides their low density (1.7 g/cm3), feature other advantages, as well, such as good ductility, the improved suppression of noise and vibrations as compared to aluminium as well as excellent castability, high size and shape stability, little shrinkage, low density combined with high strength and a low mass. They are also recyclable, thus processed alloys with their quality and properties very similar to the originally cast alloys can be obtained so that the materials can be used instead of the newly produced magnesium alloys for less important structures. A lower mass and very high strength allow for the production of parts made of this material by casting, by plastic deformation, mechanical treatment or welding. The advantages of casting magnesium alloys in conjunction with the promising outcomes of laser surface treatment investigations have set a basis for undertaking detailed scientific and research works to identify the influence of laser treatment on the structure and properties of the surface layer in casting magnesium alloys [12-15,18].

The purpose of this study is a comparative analysis of specific technologies of the laser remelting and cladding of casting magnesium alloys of MCMgAl12Zn1, MCMgAl9Zn1, MCMgAl6Zn1, MCMgAl3Zn1 using the carbide powders of TiC, WC, VC, SiC and Al_2O_3 oxide. The type of powder deposited onto the substrate was used as a criterion of technology classification, thus distinguishing between five specific technologies subjected to materials science foresight investigations. The subject of the comparative analysis are the outcomes of investigations into the structure and properties of the analysed materials, performed using specialised research apparatuses, as well as the value of the individual technologies, determined through expert studies according to the custom methodology [9], in relation to the environment as well as the long term development prospects of the technologies together with the recommended actions strategies and with the forecast multi variant development tracks. The relevance and adequacy of the assessments performed is ensured by the synergic interaction and cross supplementation of the materials science research and foresight methods. The paper also presents the outcomes of foresight research, based on reference data, [8] pertaining to the position of laser technologies in surface engineering, including laser remelting and alloying/cladding. Technology roadmaps, being a comparative analysis tool especially helpful for the small and medium sized enterprises lacking funds for conducting own research in this domain, were established at the last stage of the efforts. The results of the foresight and materials science research presented in this article are part of a broader research project [8, 17] aimed at selecting the priority innovative technologies of materials surface engineering and setting their directions of strategic development, as discussed in a series of publications, *inter alia* [18-24].

2. Materials and research methodology

The research performed is of an interdisciplinary character. The research methodology applied concerns predominantly surface engineering, being part of widely understood material engineering and technology foresight. In turn, technology foresight lies within the domain of the field of science known as organisation and management. The subject of the comparative analysis performed includes, on one hand, the results of investigations into the structure and properties of casting magnesium alloys treated using the high capacity diode laser, encompassing notably: light and scanning microscopy, X-ray phase quality analysis and an analysis of surface distribution of alloy elements as well as investigations into the properties of mechanical properties, including: hardness, microhardness and roughness. On the other hand, the long term development prospects of the individual technologies together with the recommended management strategies and the forecast multi variant development tracks are determined according to the results of the expert studies with roadmaps and the technology information sheets have been developed for them. The following five homogenous groups were distinguished between from among the technologies analysed for the purpose of experimental and comparative works by adopting, as a criterion of grouping, the type of powder deposited onto the substrate, encompassing respectively:

a. casting magnesium alloys Mg-Al-Zn undergoing laser treatment with titanium carbide TiC,
b. casting magnesium alloys Mg-Al-Zn undergoing laser treatment with tungsten carbide WC,
c. casting magnesium alloys Mg-Al-Zn undergoing laser treatment with vanadium carbide VC,
d. casting magnesium alloys Mg-Al-Zn undergoing laser treatment with silicon carbide SiC,
e. casting magnesium alloys Mg-Al-Zn undergoing laser treatment with aluminium carbide Al_2O_3.

Scientific research have been carried out on test pieces of MCMgAl12Zn1, MCMgAl9Zn, MCMgAl6Zn1, MCMgAl3Zn magnesium alloys in as-cast, after heat and laser treatment states The chemical compositions of the investigated materials are given in Table 1.

The mass concentration of main elements, %							
No.	Al	Zn	Mn	Si	Fe	Mg	Rest
1	12.1	0.62	0.17	0.047	0.013	86.96	0.0985
2	9.09	0.77	0.21	0.037	0.011	89.79	0.0915
3	5.92	0.49	0.15	0.037	0.007	93.33	0.0613
4	2.96	0.23	0.09	0.029	0.006	96.65	0.0361

Table 1. Chemical composition of investigated alloys, %

A casting cycle of alloys has been carried out in an induction crucible furnace using a protective salt bath Flux 12 equipped with two ceramic filters at the melting temperature of 750±10ºC, suitable for the manufactured material. In order to maintain a metallurgical purity of the melting metal, a refining with a neutral gas with the industrial name of Emgesalem Flux 12 has been carried out. The material has been cast in dies with betonite binder because of its excellent sorption properties and shaped into plates of 250x150x25. The cast alloys have been heated in an electrical vacuum furnace Classic 0816 Vak in a protective argon atmosphere, next MCMgAl12Zn1, MCMgAl9Zn, MCMgAl6Zn1, MCMgAl3ZMCMgAl6Zn1 magnesium alloys were used as substrate materials to laser surface treatment using high power diode laser. Laser surface alloying was conducted by remelting MCMgAl12Zn1, MCMgAl9Zn, MCMgAl6Zn1, MCMgAl3Zn surface and feeding of hard carbide particles and oxide aluminium. The alloying materials were TiC, SiC, WC, VC, Al_2O_3 powders. The powders was supplied by side injection rate of 7±1 g/min (for WC, TiC, VC powders) and 8÷9 g/min for SiC particles, Al_2O_3-4÷5 g/min.

The laser alloying was performed by high power laser diode HPDL Rofin DL 020 under an argon shielding gas. Argon was used during laser remelting to prevent oxidation of the coating and the substrate. The parameters of laser are presented in Table 2. The process parameters during the present investigation were: laser power–1.2÷1.6 kW, scan rate-0.5÷1.0 m/min.

Parameter	Value
Laser wave length, nm	940±5
Focus length of the laser beam, mm	82/32
Power density range of the laser beam in the focus plane [kW/cm^2]	0.8÷36.5
Dimensions of the laser beam focus, mm	1.8x6.8

Table 2. HPDL parameters

The observations of the investigated cast materials have been made on the light microscope LEICA MEF4A. Microstructure investigation was performed using scanning electron microscope (SEM) ZEISS Supra 25. For microstructure evaluation the Secondary Electrons (SE) detection was used, with the accelerating voltage of 5÷25 KV. Qualitative and quantitative chemical composition analysis in micro areas of the investigated coatings was performed using the X-Ray microanalysis (EDS) by mind of the spectrometer EDS LINK ISIS supplied by Oxrord.

Hardness tests were performed using Zwick ZHR 4150 TK hardness tester in the HRF scale. Tensile strength tests were made using Zwick Z100 testing machine.

In order to verify the correctness of the experimental values of hardness after laser cladding of Mg-Al-Zn casting magnesium alloys model uses a designed neural network, constructed on the basis of experimental data: the kind of used powder, the concentration of aluminium in the alloy, the laser power and speed of alloying – as the input variable – and HRF-hardness as the output variable, was used. The data set was divided into three subsets: learning (48 cases), validation (23 cases) and test (24 cases) ones. The fundamentals of the assessment of the network quality were the three characteristics of regression: average absolute error, the quotient of standard deviations, and Pearson's correlation coefficient. The quotient of the standard deviation is a gauge of the model quality used to solve regression problems. It is determined by dividing the standard deviation of prediction error and standard deviation of the output variable. A smaller value indicates a better gauge of the quality of prediction, because the smaller it is, the larger the variance explained by the model is. As a result of design and optimization of selected one way network MLP (multi layer perception) with 4 neurons in input layer – corresponding to the input variable: the nature of the powder (nominal variable), the concentration of aluminium in the alloy, the laser power and speed of alloying (numerical variables) and one numerical output variable (hardness HRF) were selected. For a nominal input variables conversion technique of one of Zn was used, while for numerical input variables and output variable the technique of conversion of variable minimax was used. The number of layers of the network was identified as three layers with two neurons in the hidden layer. The activation function in the input and output layers was defined as a linear with saturation, and in the hidden layer as the logistics, but for all the layers PSP linear functions were used. Networks were taught by methods of backpropagation of errors (50 epochs learners) and conjugate gradients (62 students ages). On the basis of achieved indicators to assess the quality of the neural network i.e., Pearson's correlation coefficients for a set of test between the calculated and actual values of output: 0.90 in the training set, 0.90 in the validation set and 0.89 in the test set, and the quotient of standard deviations for the training and test sets: <0.47 one can be infer about the accuracy in predicting the value of the output network (HRF hardness).

The reference data gathered when implementing the FORSURF project [8] was used in order to determine the strategic position of laser technologies in relation to materials surface engineering as well as the position of laser remelting and alloying/cladding in relation to surface engineering laser technologies [8]. The investigations were carried out with the three iterations of the Delphi method according to the idea of e-foresight [25] using information technology including a virtual organisation, web platform and neural networks. The five specific technologies of the remelting and cladding of casting magnesium alloys using the carbide powders of TiC, WC, VC, SiC and Al_2O_3 oxide analysed in this article were evaluated based on the opinions of key experts using the custom foresight and materials science research methodology [16]. A universal scale of relative states, being a single pole positive scale without zero, where 1 is a minimum rate and 10 an extraordinarily high rate, was used in the research undertaken. A strategic position of the relevant technologies is presented graphically with a matrix of strategies for technologies consisting of sixteen fields into which strategic development tracks were entered presenting a vision, comprised of several variants, of the future events for a 20 year timeframe according to the time intervals of 2015, 2020, 2025 and 2030. The matrix of strategies for technologies presents graphically a position of each technology group according to its value and environment influence intensity and identifies a recommended action strategy. This matrix incorporates the results of expert research, transformed with software, visualised by means of two other matrices: dendrological and meteorological matrix. The methodological structure of the both matrices is referring to the portfolio methods commonly known in management sciences, and first of all to

BCG [26] matrices enjoying their unparalleled popularity due to a reference to simple associations and intuitive reasoning, becoming an inspiration when elaborating methodological assumptions for the both matrices. A four field dendrological matrix of technology value includes the expert assessments for the relevant technologies according to the potential being the actual objective value of the specific technology group and attractiveness reflecting the subjective perception of the relevant technology group by its potential users. Depending on the potential value and attractiveness level determined in an expert assessment, each of the analysed technologies is placed into one of the matrix quarters. The wide stretching oak is the most promising quarter guaranteeing the future success in which technologies are placed characterised by a high potential and high attractiveness. The soaring cypress characterises the technologies with high attractiveness and a limited potential, and the rooted dwarf mountain pine the technologies with a large potential and limited attractiveness likely to ensure a robust position provided an appropriate strategy is applied. The least promising technologies are placed in the quarter called quaking aspen with their future success having small probability or being impossible. A four field matrix of environment influence presents, in a graphical manner, the results of how the external positive (opportunities) and negative (difficulties) factors impact the technologies analysed. Each of the technologies evaluated by the experts is placed into one of the following matrix quarters. Sunny spring illustrates the most advantageous external situation guaranteeing the future success. Rainy autumn, offering a chance for steady progress, corresponds to a neutral environment, and hot summer symbolises a stormy environment where the technology success is risky but feasible. Frosty winter informs that technology development is difficult or impossible. The results of the foresight materials science research were represented by reference data according to which a series of roadmaps for the analysed laser treatment technologies of casting magnesium alloys were established. The technology roadmaps developed with a custom concept are a convenient tool of a comparative analysis enabling to select the technologies or a group of technologies which is best in terms of the specified criterion and technology information sheets are supplementing them in technical terms.

3. Casting magnesium alloys properties dependent on technological conditions

The shape of the lase tray of the MCMgAl12Zn1, MCMgAl9Zn1, MCMgAl6Zn1, MCMgAl3Zn1 magnesium cast alloys after laser alloying with carbides and aluminium oxide using high power diode laser HPDL is presented on figures. It was found a clear influence of process parameters, in particular the laser power and the used ceramic powder on the laser tray shape and surface topography. The laser tray face after using TiC and WC powders with the feeding technique, has a regular, flat surface (Figure 1,2). In case of vanadium carbide the laser tray surface obtained after alloying is characterised by a flat shape of the remelting area, but with visible discontinuities occurred in the surface layer. Figures 3 and 4 show exemplary laser tray faces after applying the technique with putting on of the ceramic powder paste (two steps process: powder, which was mixed with a binder in form of soda glass or polyvinyl alcohol, placed on the sample surface, and following alloying with laser beam). This technique was not used in the further series of tests due to numerous discontinuities occurring on the remelted surface. The investigated laser treated materials using the powder feeding technique with SiC particles are characterised by a bulge sample surface reching above the substrate material, due to mixing of the alloyed ceramic powder with the substrate material (Fig. 5). The surface layer obtained after the process of aluminium oxide alloying is characterised by occurrence of a small caving in the middle of the laser tray surface for 2.0 kW laser power (Fig. 6). Performed investigations show, that the increase of laser power at a constant laser scanning rate influences the size of the area, where structural changes in the surface layer of the Al-Mg-Zn alloys occurs. The laser power is also related to the formation of the remelting zone bottom as well the convexity of the laser tray face, which are strongly influenced by the movements of the liquid metal.

Figure 1. View of the MCMgAl9Zn1 casting magnesium alloy face of weld after laser treatment with TiC, scan rate: 0.75 m/min, laser power: 1.2 kW

Figure 2. View of the MCMgAl12Zn1 casting magnesium alloy face of weld after laser treatment with WC, scan rate: 0.75 m/min, laser power: 1.2 kW

Figure 3. View of the MCMgAl9Zn1 casting magnesium alloy face of weld after laser treatment with WC, scan rate: 1.0 m/min, laser power: 2.0 kW

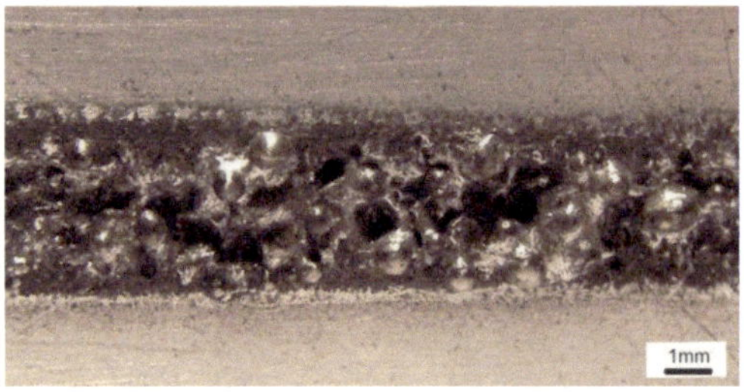

Figure 4. View of the MCMgAl9Zn1 casting magnesium alloy face of weld after laser treatment with TiC, scan rate: 1.0 m/min, laser power: 2.0 kW

Figure 5. View of the MCMgAl9Zn1 casting magnesium alloy face of weld after laser treatment with SiC, scan rate: 0.75 m/min, laser power: 2.0 kW

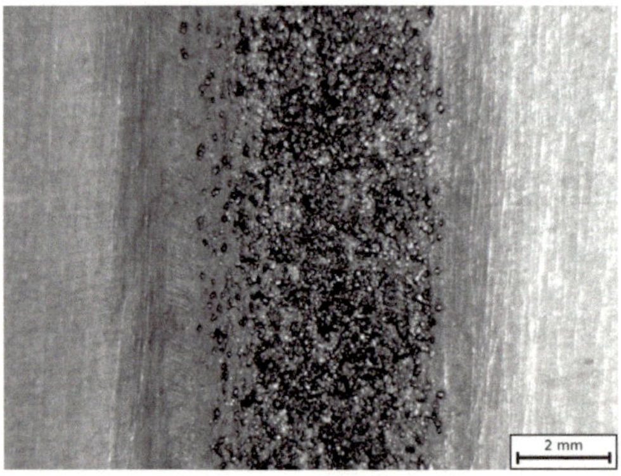

Figure 6. View of the MCMgAl9Zn1 casting magnesium alloy face of weld after laser treatment with Al_2O_3, scan rate: 0.5 m/min, laser power: 2.0 kW

Metallographic investigations results indicate, that the structure of the solidifying material after laser remelting is characterised by occurrence of areas showing a different morphology related to the crystallisation of the investigated magnesium alloys (Fig. 7-14). As a result of laser alloying there is created a defect free structure with clear grain refinement. The structure of the laser modified layer consists mainly of dispersed particles of the TiC, WC,

VC, SiC, Al_2O_3 powder placed in the Mg-Al-Zn alloy matrix. Morphology of the remelted zone after laser alloying consists mainly of dendrites with $Mg_{17}Al_{12}$ plate like eutectic and Mg occurred in the interdendritic areas, whose main axes are oriented along the heat transport directions. Moreover the morphology of the composite structure of the area after laser alloying results from the change of the hypo eutectic alloy to an hyper eutectic one, depending on the dissolution and distribution of the ceramic powder used and process parameters applied for the surface layer treatment.

Investigations carried out using the scanning electron microscope have confirmed the presence of zonal structure in the surface layer of the investigated magnesium cast alloys (Figure 7-12). In the remelted zone there occurs a dendritic structure, coming into existence according to the heat transport direction. The dendritic structure occurs together with not dissolved particles of the used carbide or aluminium oxide powder (Fig. 13,14). Morphology of the area after laser alloying, as well the amount and distribution of carbide particles depends on the applied laser parameters. As a result of metallographic investigations of the MCMgAl3Zn1, MCMgAl6Zn1, MCMgAl9Zn1 and MCMgAl12Zn1 alloy there was found evenly distributed particles over the remelting zone (Figure 7-12). In the upper area of the remelting zone in which vanadium carbide was alloyed some turbulences can be seen, which are caused by the convective movement of the melt and the ceramic powder during the remelting process. Chemical composition investigations using energy dispersive X-ray spectrometer (EDS), as well as investigation of surface distribution of the chemical elements carried out on a cross section of the surface layer of the cast magnesium alloy Mg-Al-Zn using TiC, WC, VC, SiC, Al_2O_3 powders confirms the occurrence of magnesium, aluminium, zinc, manganese, coal, and also respectively titanium, tungsten, vanadium, silicon and oxygen in the laser modified layer, and indicate a lack of solubility of the alloyed particles (Fig. 11-14).

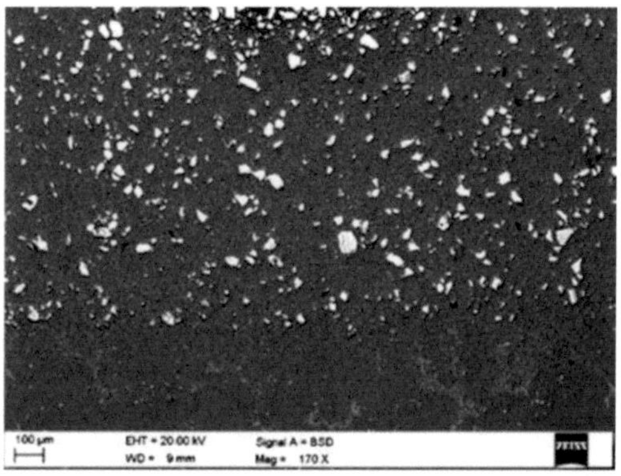

Figure 7. Scanning electron microscope micrograph of cross section laser modified surface of the MCMgAl9Zn1 alloy with TiC (laser power 1.6 kW), scan rate: 0.75 [m/min]

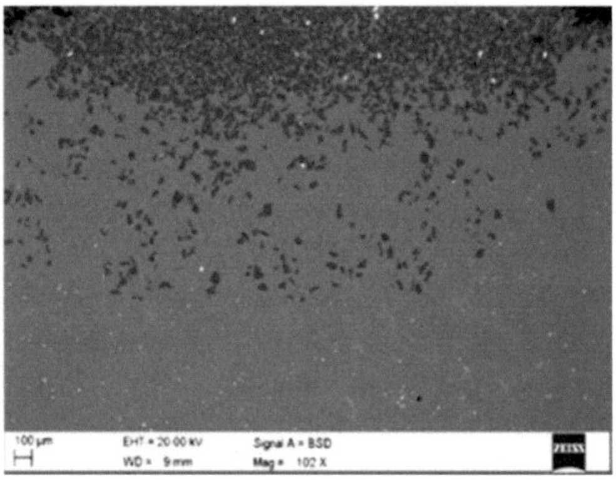

Figure 8. Scanning electron microscope micrograph of cross section laser modified surface of the MCMgAl9Zn1 alloy with SiC (laser power 2.0 kW), scan rate: 0.75 [m/min]

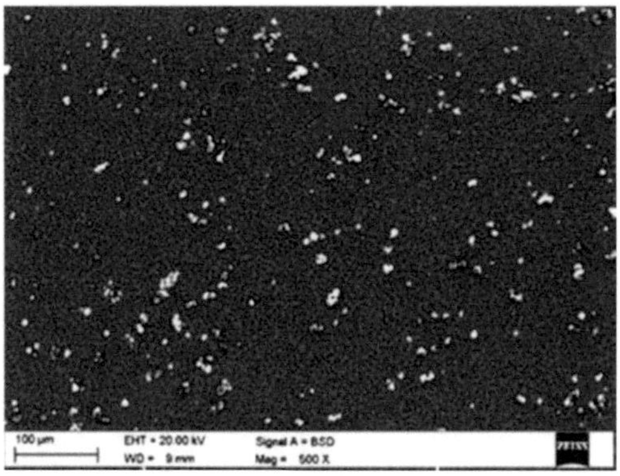

Figure 9. Scanning electron microscope micrograph of laser modified surface of MCMgAl12Zn1 alloy with WC particles of the central modified zone, scan rate: 0.75 [m/min], laser power: 2.0 [kW]

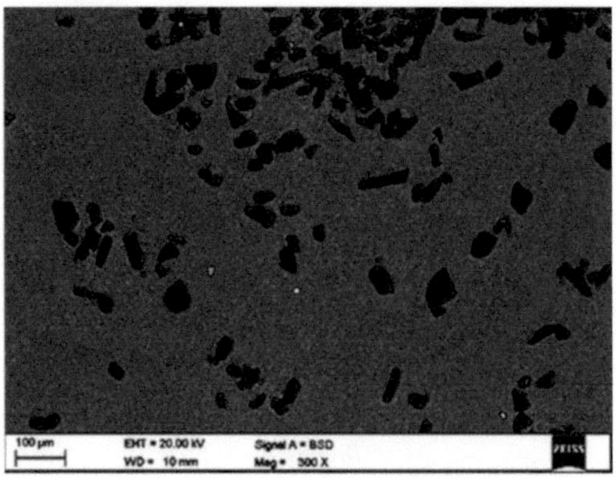

Figure 10. Scanning electron microscope micrograph of laser modified surface of MCMgAl6Zn1 alloy with SiC particles of the central modified zone, scan rate: 0.75 [m/min], laser power: 2.0 [kW]

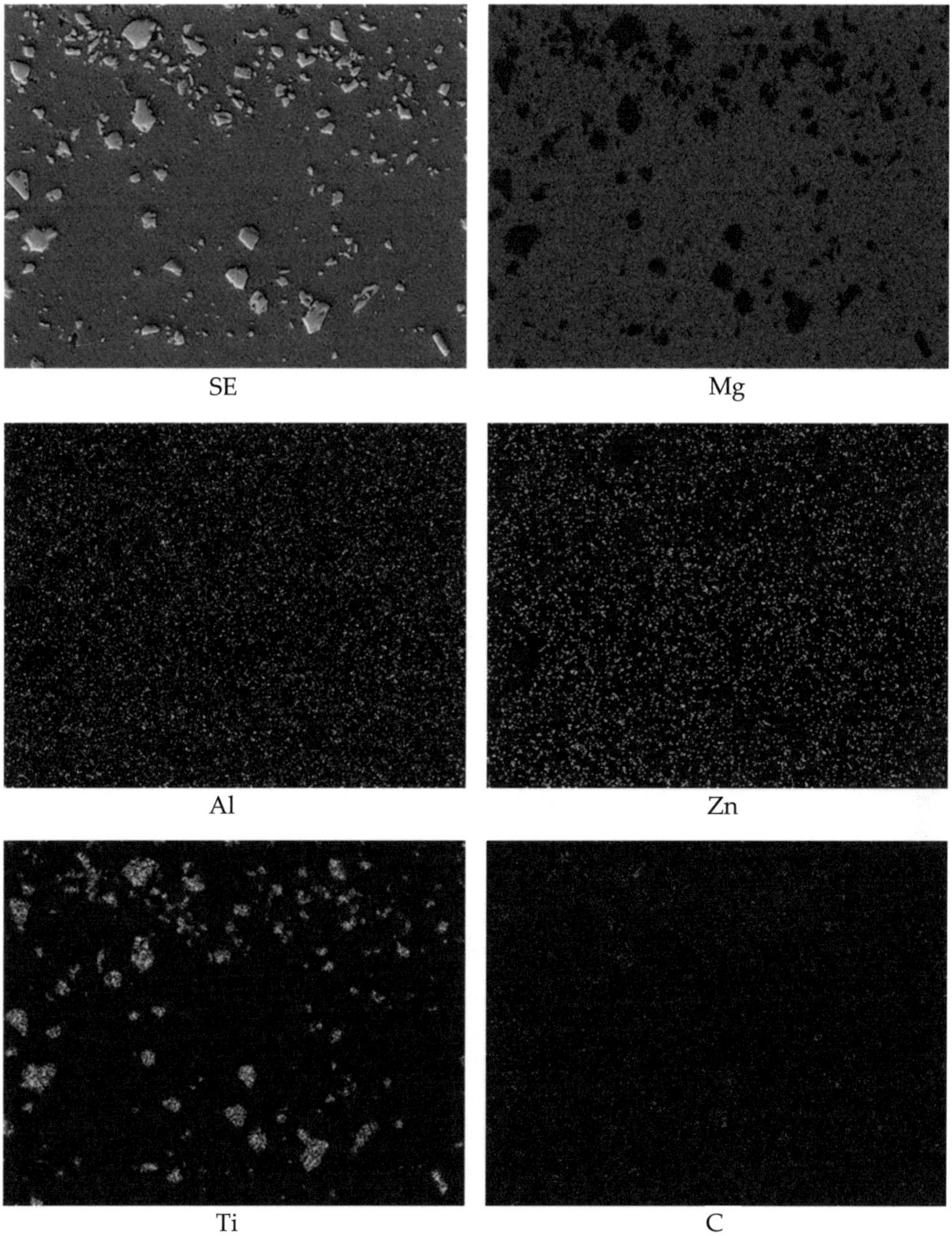

Figure 11. The area analysis of chemical elements alloy MCMgAl6Zn1 after laser treatment with TiC, scan rate: 0.75 [m/min], laser power: 1.6 [kW]: image of the secondary electrons (SE) and maps of elements' distribution

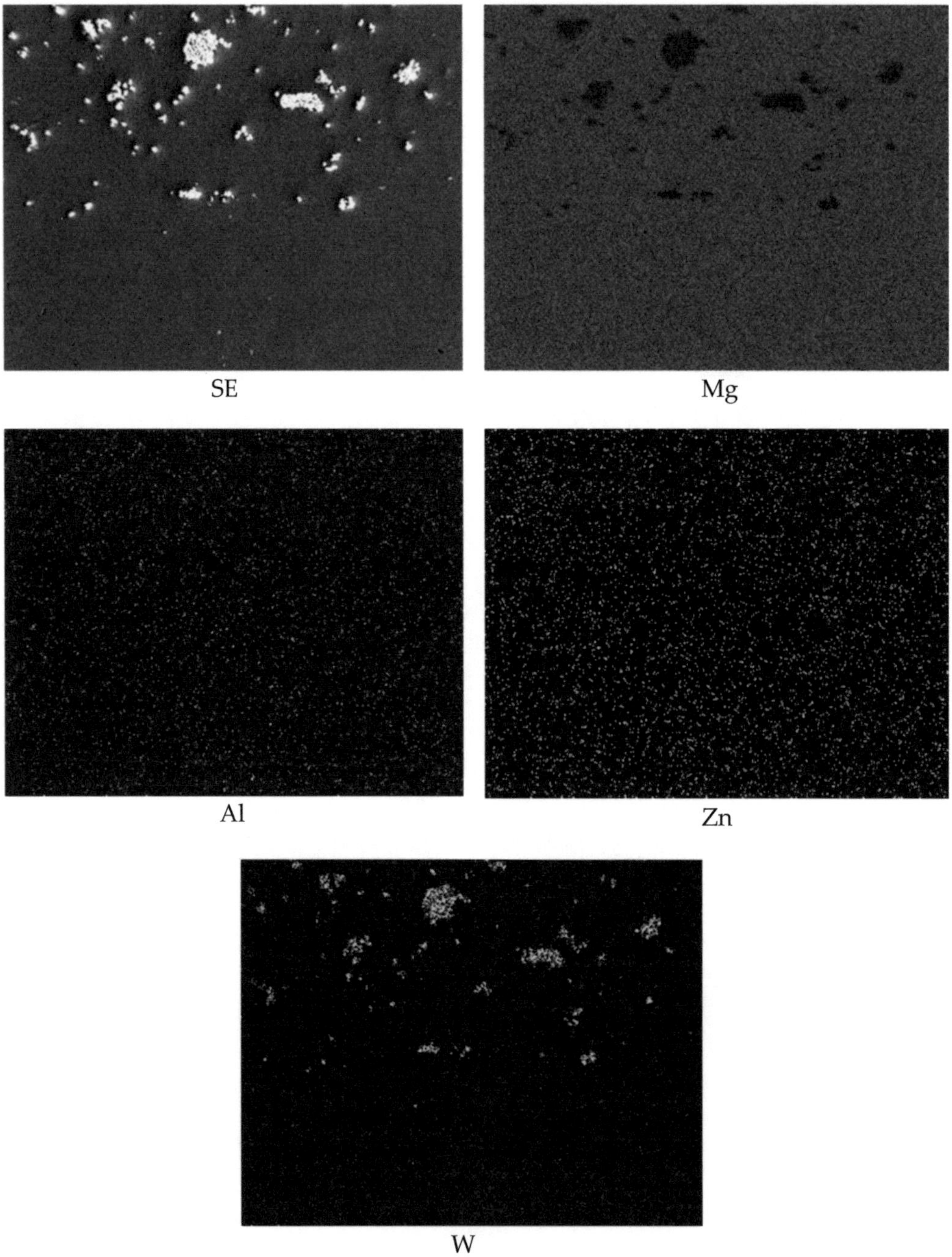

Figure 12. The area analysis of chemical elements alloy MCMgAl6Zn1 after laser treatment with WC, scan rate: 0.75 [m/min], laser power: 2.0 [kW]: image of the secondary electrons (SE) and maps of elements' distribution

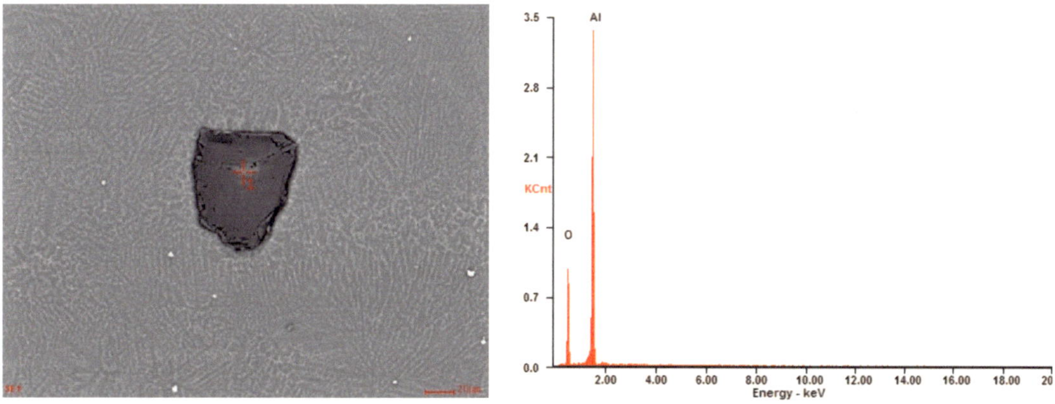

(a) (b)

Figure 13. Structure of the laser modified surface of MCMgAl9Zn1 alloy with Al_2O_3 particles of the central modified zone, scan rate: 0.5 [m/min], laser power: 2.0 [kW], a) SEM micrograph, b) EDS microanalysis of the Al_2O_3 particles with surface layer in point 1 marked on Fig. 13 a

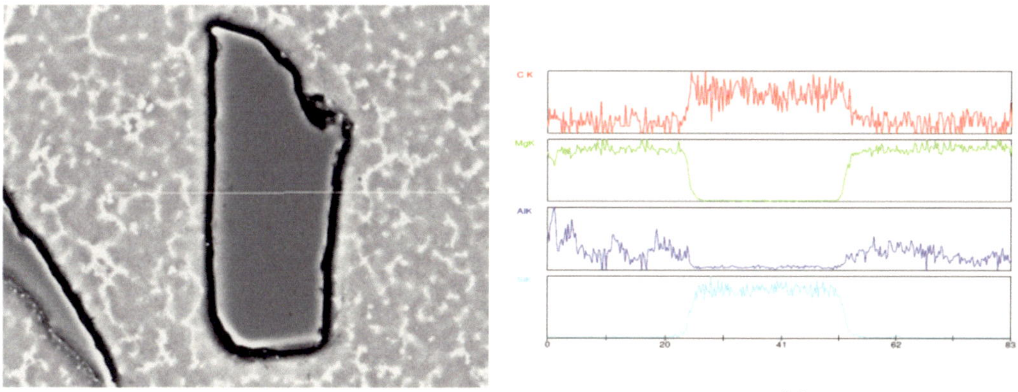

(a) (b)

Figure 14. Scanning electron microscopy micrograph of MCMgAl9Zn1 alloy after laser alloying with SiC particles, laser power: 2.0 kW, scan rate: 0.75 m/min, powder feed rate: 0.75 m/min, a) SEM micrograph, b) linear analysis of the chemical composition changes marked on Fig. 14 a

Hardness measurements results of the Mg-Al-Zn cast magnesium alloy after remelting and alloying with WC, TiC, VC, NbC, SiC carbides and Al_2O_3 oxide (Figure 15) show, that in most cases laser treatment of the surface layer causes an hardness increase. The highest hardness increase of 56 HRF compared to the hardness results achieved for the material after standard heat treatment was obtained for the MCMgAl3Zn1 alloy alloyed with TiC powder with laser power of 1.2 kW and laser scanning rate of 1.0 m/min. For the MCMgAl6Zn1 alloy the highest hardness (93.4 HRF) after laser treatment was measured for the material alloyed with TiC powder with laser power of 1.2 kW and laser canning rate of 0.75 m / min.

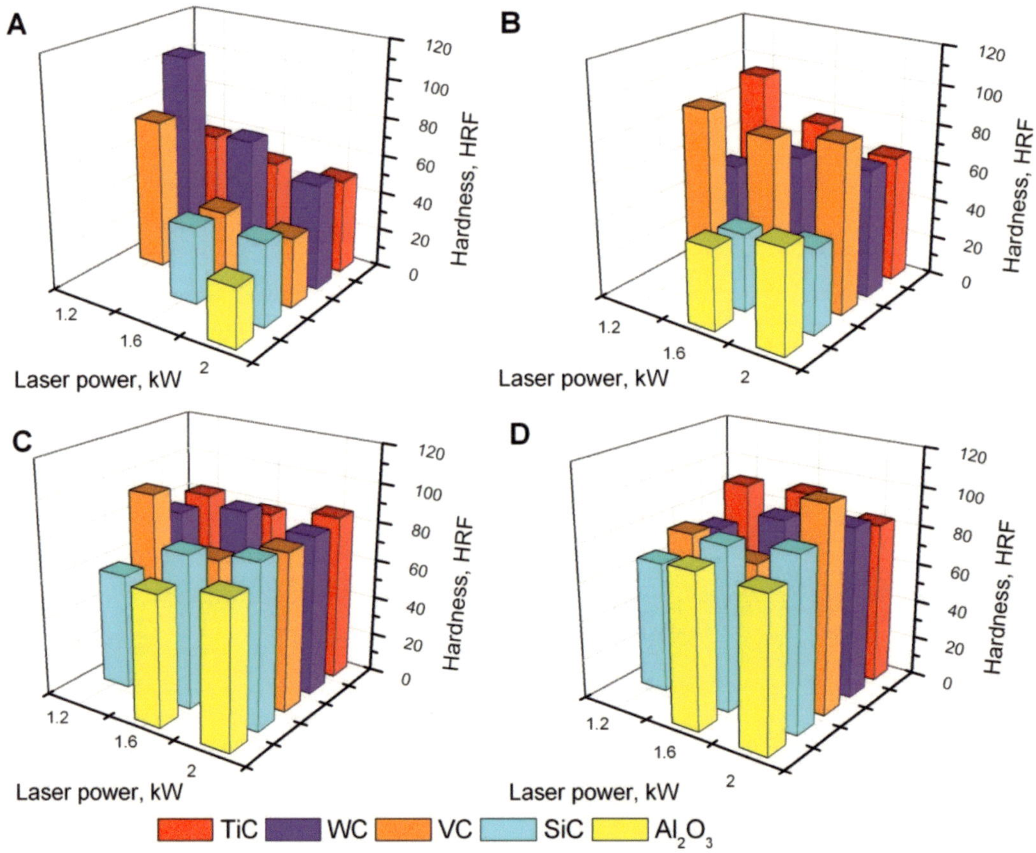

Figure 15. Change in the average hardness of the surface layer of casting magnesium alloys after laser treatment: A – MCMgAl3Zn1, B – MCMgAl6Zn, C – MCMgAl9Zn1, D – MCMgAl12Zn1

Furthermore, on the basis of the outworked neural network model diagrams of the impact of laser power, concentration of aluminium, and also the type of powder on the hardness of the analyzed casting magnesium alloys after laser treatment of the surface layer (Figs. 16) were made. The diagrams in most cases concern the remelting speed of 0.75 m/min, corresponding to the optimum geometry of the path of the laser. The obtained results clearly show that MCMgAl12Zn1 casting magnesium alloys alloyed by TiC and WC powders with a laser power of 2.0 kW and a speed of 0.75 m/min. are characterised by the highest hardness.

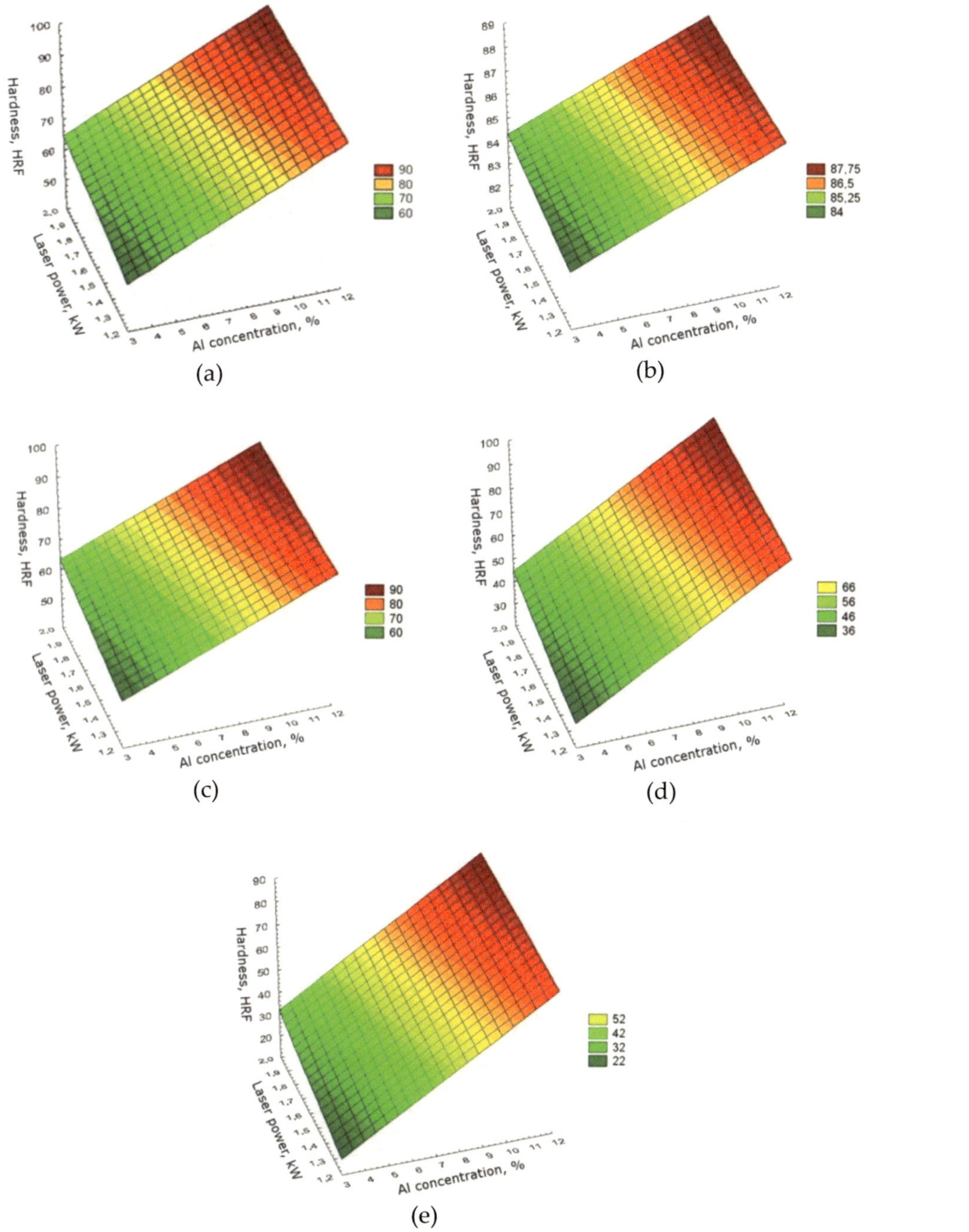

Figure 16. Simulation of the laser power and aluminium concentration (wt. %) influence on hardness of casting magnesium alloys after laser alloyed with: a) TiC, b) VC, c) WC, d) SiC, e) Al_2O_3particles, scan rate 0.75 m/min (a-d), scan rate 0. 5 m/min (e)

4. Forecasted development of laser technologies in surface engineering

The reference data gathered when implementing the FORSURF project for surface properties formation leading technologies of engineering materials and biomaterials was used in order to determine the strategic position of laser technologies in relation to materials surface engineering as well as the position of laser remelting and alloying/cladding in relation to surface engineering laser technologies [8]. Over 300 independent experts from many countries representing scientific, business and public administration circles have taken part in the FORSURF technology foresight. The experts have completed approx. 650 multi question surveys and held thematic discussions during 10 Workshops. A collection of 140 critical technologies, 10 for each thematic group, was selected for the above 14 thematic groups from the initially inventoried approx. 500 specific technology groups. The scientific and research methods of evaluating the state of the art for a particular concept, technology review and a strategic analysis with integrated methods were used for this purpose, including: extrapolation of trends, environment scanning, STEEP analysis, SWOT analysis, expert panels, brainstorming, benchmarking, multi criteria analysis, computer simulations and modelling, econometric and static analysis. Next, the technologies were thoroughly analysed with three iterations of the Delphi method carried out in consistency with the idea of e-foresight using information technology encompassing a virtual organisation, web platform and neural networks, with a universal scale of relative states being a single pole positive scale without zero, where 1 is a minimum rate and 10 an extraordinarily high rate.

Foresight investigations with the sample size of 198 have revealed a very robust strategic position of laser technologies among other materials surface engineering technologies. The experts found that that laser technologies have the best industrial application prospects in the group of all the analysed materials surface engineering technologies in the nearest 20 years. 78% of the surveyed held such a view. Nearly a three fourth of the respondents (73%) maintain that numerous scientific and research studies will be devoted to such technologies in the analysed time horizon. 70% of the persons surveyed claim that the thematic area of "Laser technologies in surface engineering" is crucial and its importance should be absolutely rising so that an optimistic scenario can come true of the country's/Europe/World development, i.e. "Race won" assuming that the potential available is adequately utilised to fulfil the strategic objectives of development and so that people, statistically, are better off, social attitudes are optimistic and the prospects for the coming years bright. 81% of the surveyed persons argue that the significance of laser technologies in relation to other materials surface engineering technologies will be growing, whereas 18% maintain it will remain on the same level with only 3 individuals asserting that the role will diminish over the next 20 years. The excellent results of technology foresight elaborated based on the reference data point to, therefore, the anticipated key role of laser technologies for the advancement of the overall materials surface engineering (mezo scale) and for the development of the entire domestic/European/global economy (macro scale) [23]. The results of the foresight research discussed, presenting the position of laser technologies against materials surface engineering as a whole, are provided on Fig.17.

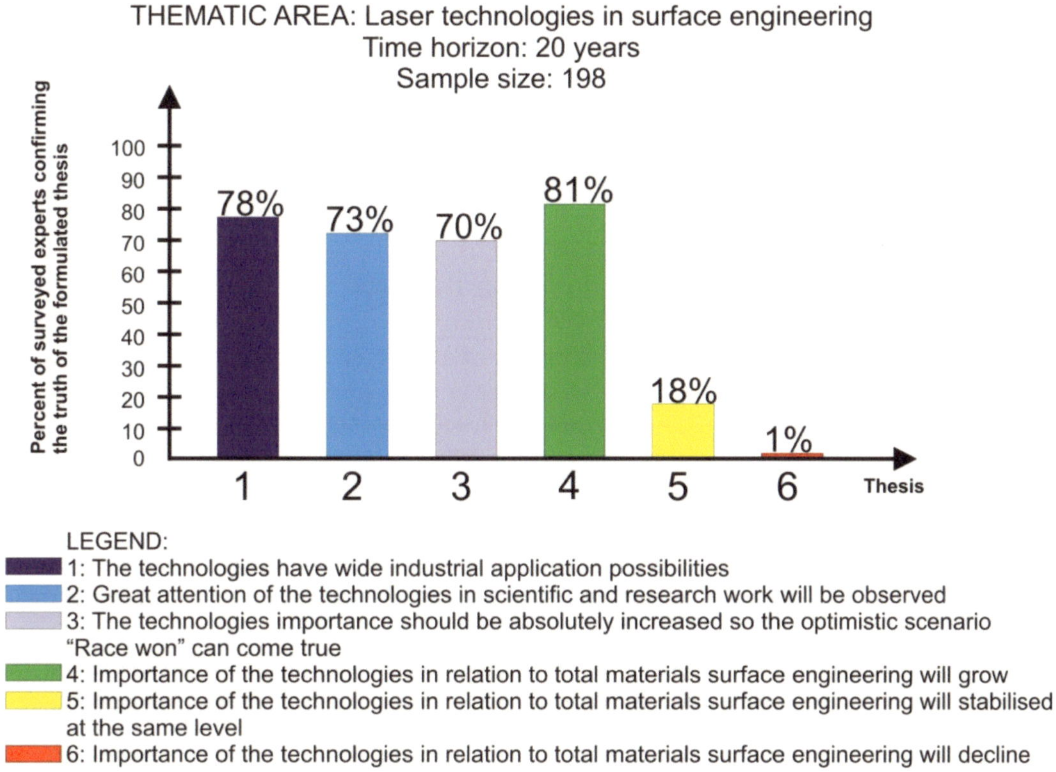

Figure 17. The position of laser technologies against materials surface engineering as a whole [30]

The position of laser remelting and alloying/cladding vis-à-vis other laser technologies in surface engineering has undergone a thorough foresight study made by the selected key experts, specialists in the field of laser technologies. The two technology groups analysed are characterised by stable, predictable development prospects. 40 % of the experts surveyed maintain that the technology group of laser alloying/cladding, characterised by its late maturity, falls within the group of critical technologies and its importance should be absolutely rising so that an optimistic scenario of the country's/Europe/World development, i.e. "Race Won" comes true in the nearest 20 years. 20% of the experts attending the study held a similar view for the base technology of laser remelting.

5. Laser surface treatment of casting magnesium alloys in the future

The long run development prospects were identified based on the materials science experiments and expert studies performed by means of the custom methodology [9] for the individual groups of specific technologies, including the laser treatment of casting magnesium alloys, i.e. respectively: (A) titanium carbide TiC, (B) tungsten carbide WC, (C) vanadium carbide VC, (D) silicon carbide SiC and (E) aluminium oxide Al_2O_3. The recommended action strategies and the predicted multi variant development tracks and technology roadmaps were also developed and information sheets were prepared.

As part of the research conducted, the key experts in the first place assessed the analysed technology groups with a universal scale of relatives states consisting of ten points (max: 10, min: 1) for their attractiveness and potential and the result obtained was entered into the dendrological matrix of technology value [18]. The analysis made has shown that all the groups of technologies were classified to the most promising quarter called wide stretching oak, encompassing the technologies with a high potential and attractiveness. The best score of A (9,65; 9,75) was attained for casting magnesium alloys undergoing laser treatment with titanium carbide, and the lowest score of D (7,55; 8,45) was seen for those where silicon carbides were used for laser treatment. Positive and negative environment influence on the relevant technology groups was evaluated with a meteorological matrix of environment influence. The results of a multi criteria analysis of the experts' scores acquired in the survey taking process were entered into the matrix [18]. The results of the studies made reveal that the environment for all the technology groups subjected to the studies is unusually favourable, bringing multiple opportunities and few difficulties. Hence, all the analysed technology groups were found in the quarter corresponding to sunny spring, boding very well for their development. Again, the technology group referred to as A (4,04; 7,36) scored highest, and the lowest score was given to the technology group called E (3,77;6,02). The results of the studies presented graphically with the dendrological and meteorological matrix were at the next stage of the scientific pursuits entered into the matrix of strategies for technologies by means of the software developed for this purpose (Fig.18). The matrix presents, in a graphical manner, a position of the relevant technology groups of the laser treatment of casting magnesium alloys with carbides and aluminium oxide according to its value and environment influence intensity and identifies an appropriate action strategy. The oak in spring strategy is recommended for all those analysed technology groups that are boding well. The strategy consists in developing, strengthening and implementing an attractive technology with a large potential in the industrial practise to achieve a spectacular success.

The next stage of the research consists of identifying the strategic development tracks for the individual technologies/technology groups according to the experts' opinions, representing a forecast of their development for the years of: 2015, 2020, 2025 and 2030 according to the three variants: optimistic, pessimistic and the most probable one. They are next visualised against the technology strategy matrix. The numerical values, being an outcome of all the investigations performed for the three analysed groups of technologies, are listed in Table 3. Due to relatively small differences between the individual analysed groups of technologies at a macro scale, the strategic development tracks established for them have a similar direction, showing minor differences and are discussed further on with a representative example of the laser treatment of casting magnesium alloys with titanium carbide TiC.

The most probable strategic development track of the leaser treatment of casting magnesium alloys with titanium carbide TiC assumes that the environment conditions shift from friendly spring to risky summer while maintaining a high potential and attractiveness characteristic for wide stretching oak. The environment will become more stable in the subsequent years transiting into the autumn phase. It is anticipated that an attractive, stable technology will become successful at the predicted market with other markets being sought

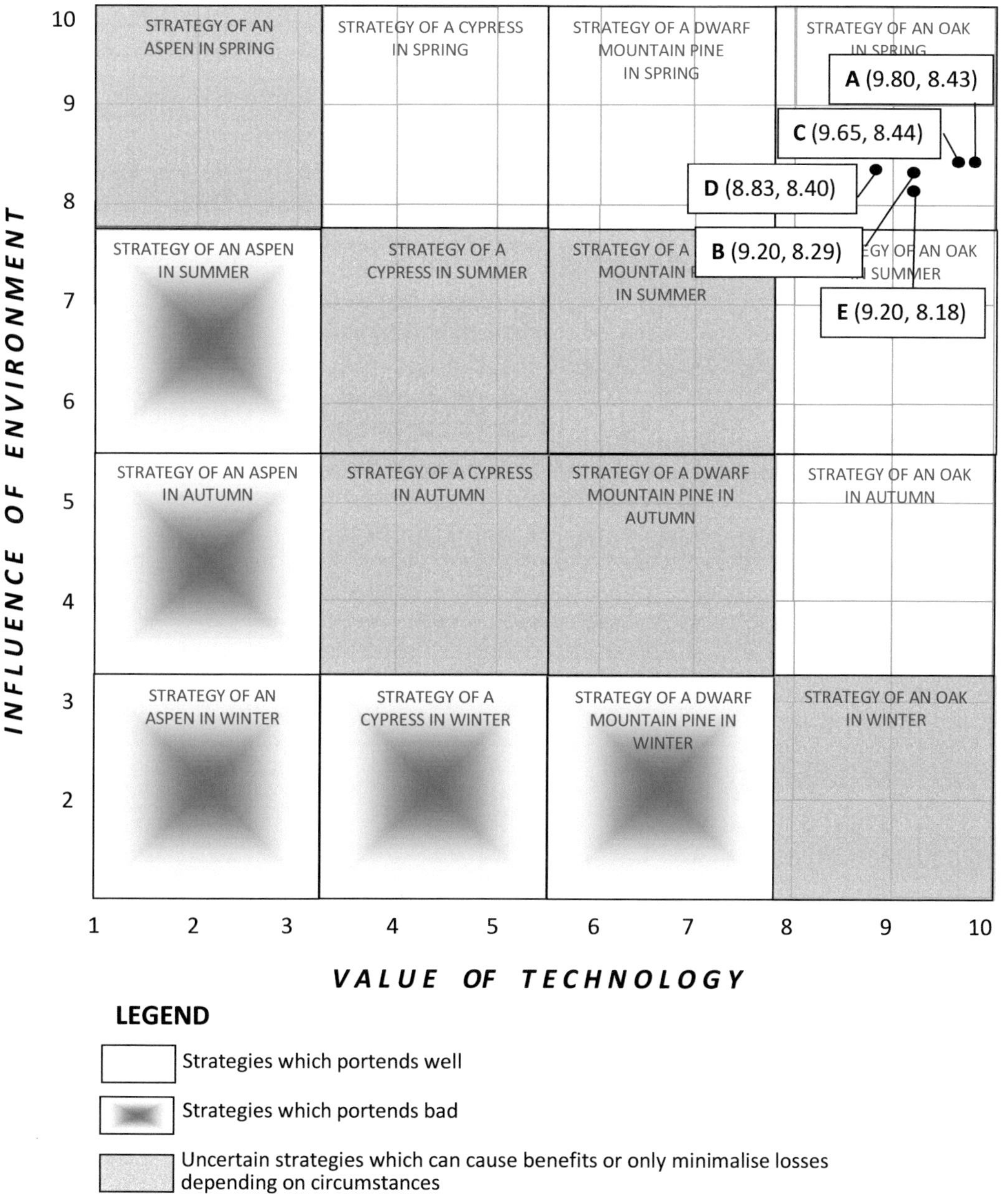

Figure 18. The matrix of strategies for technology called the laser cladding and remelting of casting magnesium alloys using TiC (A), WC (B), VC (C), SiC (D) carbide and Al2O3 oxide (E) powders [18]

Technology		Steady state 2010-11	Type of strategic development tracks	Years			
Symbol	Name			2015	2020	2025	2030
(A)	The laser treatment of TiC in the Mg-Al-Zn surface	Strategy of an oak in spring A (9.8, 8.4)	(O)	(9.8, 6.5)	(9.9, 7.0)	(9.9, 8.0)	(9.9, 9.0)
			(P)	(9.8, 6.0)	(9.8, 2.0)	(6.0, 2.0)	(3.0, 1.8)
			(MP)	(9.7, 6.0)	(9.8, 7.0)	(9.8, 4.5)	(9.9, 4.8)
(B)	The laser treatment of WC in the Mg-Al-Zn surface	Strategy of an oak in spring B (9.2, 8.3)	(O)	(9.2, 5.6)	(9.3, 6.2)	(9.4, 7.0)	(9.4, 8.0)
			(P)	(9.2, 5.3)	(9.2, 1.6)	(5.7, 1.6)	(3.0, 1.4)
			(MP)	(9.2, 5.6)	(9.2, 6.0)	(9.3, 3.9)	(9.3, 4.2)
(C)	The laser treatment of VC in the Mg-Al-Zn surface	Strategy of an oak in spring C (9.7, 8.4)	(O)	(9.7, 6.2)	(9.8, 6.5)	(9.8, 7.5)	(9.8, 8.5)
			(P)	(9.7, 5.7)	(9.7, 1.8)	(5.9, 1.8)	(3.0, 1.5)
			(MP)	(9.6, 5.7)	(9.7, 6.5)	(9.7, 4.0)	(9.8, 4.3)
(D)	The laser treatment of SiC in the Mg-Al-Zn surface	Strategy of an oak in spring D (8.8, 8.4)	(O)	(8.8, 5.6)	(8.8, 6.0)	(8.9, 7.0)	(9.0, 8.2)
			(P)	(8.8, 5.7)	(8.7, 1.7)	(5.9, 1.7)	(3.0, 1.4)
			(MP)	(8.8, 5.6)	(8.8, 5.4)	(8.8, 4.0)	(8.9, 4.3)
(E)	The laser treatment of Al_2O_3 in the Mg-Al-Zn surface	Strategy of an oak in spring E (9.2, 8.2)	(O)	(9.2, 5.6)	(9.4, 6.0)	(9.4, 7.1)	(9.4, 8.1)
			(P)	(9.2, 5.2)	(9.2, 1.5)	(5.6, 1.5)	(3.0, 1.4)
			(MP)	(9.2, 5.6)	(9.3, 6.0)	(9.3, 4.0)	(9.3, 4.1)

Table 3. Strategic development tracks of laser treatment of Mg-Al-Zn casting magnesium alloys using carbide and oxide powders. Types of strategic development tracks: (O) - optimistic, (P) - pessimistic; (MP) - the most probable [18]

for along with the new groups of potential clients and new products manufacturable with the specific technology. An optimistic laser treatment development track for casting alloys with titanium carbide TiC assumes that although a number of temporary (2015-2020) difficulties occur in the environment, the opportunities emerging at the same time can be exploited with those opportunities defining the development of this technology group in the further years ensuring their return to the friendly area of sunny spring. This, combined with the maintaining high attractiveness and technology potential, will ensure a spectacular success. A pessimistic variant defined by the third strategic development track for the technology group envisages that the global downturn would become even harsher due to the unfolding disadvantageous political and economic situation. This will cause more and more difficulties in the environment (year of 2015) and fewer and fewer opportunities making it necessary to operate in 2020 in the unfavourable conditions of frosty winter. The economic circumstances will be unfriendly, making the potential users less interested in the technology group. In 2025 the analysed technology group, being rooted dwarf mountain pine, by using a large potential representing an objectively high value of the technology, will make attempts to withstand the difficulties while regularly weakening, so that it transits to the field of quaking aspen in winter in 2030 with withdrawal from the market being then advisable.

6. Technology roadmapping

A series of roadmaps of the technology groups analysed was created on the basis of the results of experimental and comparative studies. The roadmaps serve as a comparative analysis tool permitting to select the technologies or technology groups best in terms of the criterion defined [27-29]. The roadmaps, prepared with a custom concept, have their set up corresponding to the first quarter of the Cartesian system of coordinates. The following time intervals, respectively: current situation (2010-11), goals fulfilment methods (2020) and long term objectives (2030) are provided on the axis of abscissa, i.e. time layer, concept layer, product layer, technology layer, spatial layer, staff layer and quantitative layer, made up of more detailed sub layers. The upper most layers of the technology roadmap are most general and determine the all social and economic reasons and causes of the actions taken. The middle layers are characterising a product and its manufacturing technology. The bottom layers are determining organisational and technical matters concerning the place, contractor and costs. Cause and effect relationships, capital ties, time correlations and two directional data and/or resources flow take place between the individual layers and sub layers as signified graphically with the different types of arrows. Fig.19 presents a representative technology roadmap drafted for the laser cladding of vanadium carbide VC particles into the surface of casting magnesium alloys Mg-Al-Zn. Table 4 presents an aggregate list containing the selected data being an extract from all the technology roadmaps developed for the analysed casting magnesium alloys subjected to laser treatment. The technology information sheets are detailing out and supplementing the technology roadmaps. They contain technical information very helpful in implementing a specific technology in the industrial practice, especially in SMEs lacking the capital allowing to conduct own research in this field.

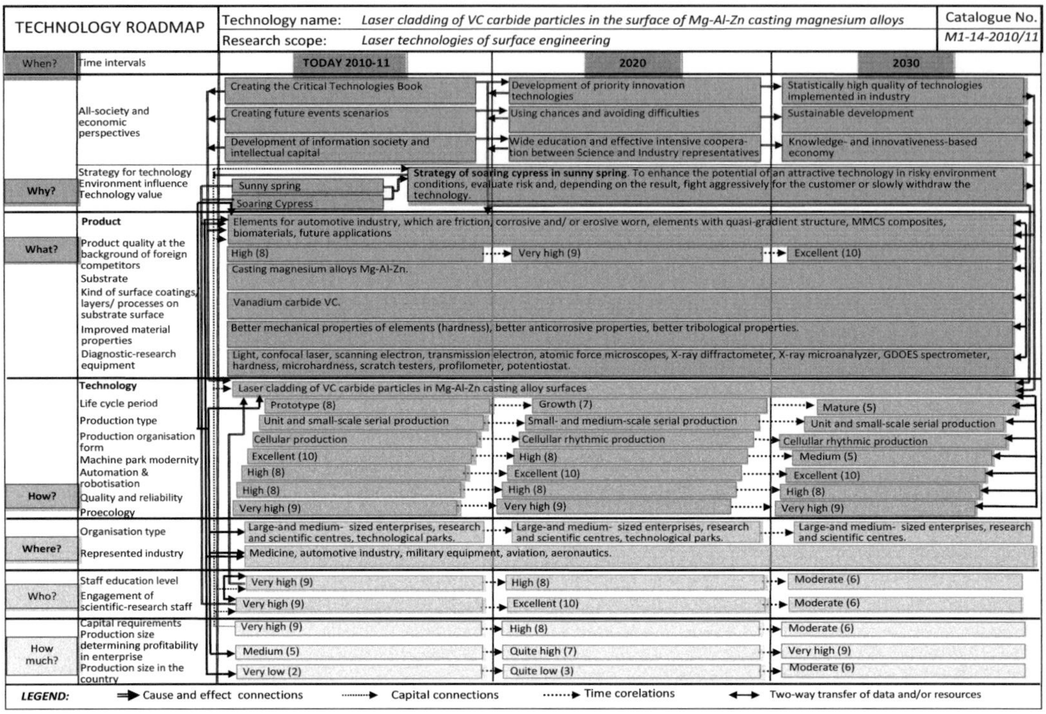

Figure 19. Demonstrating technology roadmap for laser cladding of TiC in the substrate of Mg-Al-Zn casting magnesium alloys

Technology symbol	Analysed factors																							
	(1)			(2)			(3)			(4)			(5)			(6)			(7)			(8)		
	Time horizon																							
	a	b	c	a	b	c	a	b	c	a	b	c	a	b	c	a	b	c	a	b	c	a	b	c
(A)	8	7	3	10	8	5	8	9	10	10	9	9	10	8	6	10	8	5	5	7	8	3	5	8
(B)	8	7	5	10	8	5	7	8	8	8	8	8	9	5	5	9	9	4	6	7	8	2	5	7
(C)	8	7	5	10	8	5	8	8	8	9	9	9	9	8	6	9	8	6	5	7	9	2	4	6
(D)	8	7	5	8	8	5	6	8	9	8	9	9	9	5	5	9	7	5	6	7	8	3	6	9
(E)	8	7	3	8	8	6	8	8	9	7	7	9	9	5	5	7	7	7	5	7	8	2	5	7

LEGEND

Technology symbol	**Analysed factors**	**Time horizon**
(A) The laser treatment of TiC in the Mg-Al-Zn surface	(1) Live cycle period	
	(2) Machine park modernity	a: 2010-11 years
(B) The laser treatment of WC in the Mg-Al-Zn surface	(3) Quality and reliability	
(C) The laser treatment of VC in the Mg-Al-Zn surface	(4) Proecology	b: 2020 year
	(5) Staff education level	
(D) The laser treatment of SiC in the Mg-Al-Zn surface	(6) Capital requirements	c: 2030 year
(E) The laser treatment of Al_2O_3 in the Mg-Al-Zn surface	(7) Production size determining profitability in enterprise	
	(8) Production size in the country	

Note: Research results are presented in universal scale of relative state, where:1 is minimal and 10 is excellent level.

Table 4. Selected source data used for preparation of technology roadmaps for investigated laser treated casting magnesium alloys

7. Summary

This chapter of the book presents the results of interdisciplinary foresights materials science research pertaining to five groups of specific technologies including the high capacity diode laser treatment of casting magnesium alloys with, respectively: (A) titanium carbide TiC, (B) tungsten carbide WC, (C) vanadium carbide VC, (D) silicon carbide SiC and (E) aluminium oxide Al_2O_3. Materials science investigations were carried out in particular including light and scanning microscopy, X-ray phase qualitative analysis and surface distribution analysis of alloy elements. Investigations into mechanical properties were also held including hardness, microhardness and roughness as well as expert studies. The long term development prospects of casting magnesium alloys subjected to laser treatment were identified using a custom methodology, along with the recommended action strategies and the forecast multi variant development tracks. The results of the foresight research [8] based on reference data [8] for the position of laser technologies were also presented, including remelting and laser alloying/cladding against materials surface engineering in general.

The results presented for the materials science research reveal a promising improvement in the mechanical properties of the material investigated. Laser cladding and remelting with all the carbide and oxide powders referred to above influences the refinement of structure within the entire investigated scope of laser power and also influences the varied grain size in the individual zones of the surface layer of the alloys investigated. Two zones occur in the surface layers: remelting zone (RZ) and heat affected zone (HAZ) with their characteristic values (layer thickness) depending on the laser power used and the alloying material used. The structure of the material solidifying after laser remelting is characterised by a varied morphology and consists of dispersive particles of the TiC, WC, VC, SiC carbide or Al_2O_3 oxide applied, of dendrites with the lamellar eutectic of $Mg_{17}Al_{12}$ and Mg in interdendritic areas, with their main axes being oriented towards the directions of heat evacuation and also of precipitates containing Mg and Si, as well as of phases with a high concentration of Mn and Al. In addition, a morphology of the composite structure of an alloyed area was obtained by changing a hypoeutectic to hypereutectic alloy, depending on the arrangement of the alloyed elements and by changing the surface laser treatment process parameters. In the Mg-Al-Zn casting magnesium alloys subjected to remelting and alloying with carbides and oxide, the maximum hardness of approx. 103 HRF was achieved for the MCMgAl12Zn1 alloy alloyed with titanium carbide at the laser power of 1.2 kW and at the alloying rate of 0.5 m per min., as a result of grain refinement and the occurrence of hard particles of the powders applied.

One should conclude while analysing the results obtained for the research that it is feasible to use the investigated Mg-Al-Zn alloys and their treatment technologies also in an alternative fashion for the surface layers ensuring possibly the most favourable "quasi gradient" properties on the section of the products in the industrial practise. Widespread potential applications are identified especially for the aviation and automotive industry where the following properties are required: a small mass density of products, higher wear resistance, improved strength parameters of elements as well as repairing the ready elements. The best developmental and applicational prospects stemming from an analysis of mechanical properties of the casting magnesium alloys subjected to laser treatment are exhibited by those materials into which the particles of titanium carbide TiC (technology A) and vanadium carbides VC (technology C) were cladded. An analytical tool likely to facilitate the future implementation of the technologies analysed, in particular in small and medium sized enterprises, are the roadmaps and technology information sheets, prepared at the last stage of the research works, containing concise knowledge and the results of the experimental and expert works were used for this purpose.

Author details

Anna Dobrzańska-Danikiewicz, Tomasz Tański,
Szymon Malara and Justyna Domagała-Dubiel
Faculty of Mechanical Engineering, Silesian University of Technology, Gliwice, Poland

Acknowledgement

Research were financed partially within the framework of the structural project POIG.01.01.01-00-023/08-03 FORSURF, headed by Prof. L.A. Dobrzański

8. References

[1] http://ec.europa.eu/europe2020/priorities/sustainable growth/index_pl.html

[2] http://ec.europa.eu/research/innovation union/index_en.cfm?pg=home

[3] Dreher C (2007), Manufacturing visions: A holistic view of the trends for European manufacturing, Edited by Montorio M, Taisch M, Thoben KD, Advanced Manufacturing. An ICT and Systems Perspective, Taylor & Francis Group, London. 315 p.

[4] Dosch H, Van de Voorde MH (eds.) (2009), Gennesys. White Paper. A New European Partnership between Nanomaterials Science & Nanotechnology and Synchrotron Radiation and Neuron Facilities, Max Planck Insititut für Metalforschung, Stuttgart. 498p.

[5] NANOMAT, www.nanomat.eitplus.pl (in Polish).

[6] Advanced Industrial and Ecological Technologies for Sustainable Development of Poland, www.portaltechnologii.pl/3index/index.html.

[7] Edited by Gambin B, Łojkowski W, Świderska-Środa A (2010) FOREMAT, Technology Development Scenarios of Modern Metallic, Ceramic and Composites Materials. Reports of Project Co-Operators, Unipress Publisher, Radom, 2010 (in Polish). 442 p.

[8] FORSURF, www.forsurf.pl 2009-2012 (in Polish).

[9] Horst EF, Mordike BL (2006) Magnesium Technology. Metallurgy, Design Data, Application, Springer-Verlag, Berlin Heidelberg, 2006. 707 p.

[10] Tański T, Dobrzański L A, Labisz K (2010) Investigations of microstructure and dislocations of cast magnesium alloys. Journal of Achievements in Materials and Manufacturing Engineering. 42/1-2: 94-101.

[11] Dobrzański LA, Tański T (2009) Influence of aluminium content on behaviour of magnesium cast alloys in bentonite sand mould. Solid State Phenomena. 147-149: 764-769.

[12] Dobrzański L A, Tański T, Malara Sz, Król M, Domagała-Dubiel J (2011) Contemporary forming methods of the structure and properties of cast magnesium alloys. In Czerwinski F, editor. Magnesium Alloys - Design, Processing and Properties. Rijeka: InTech. pp. 321-350.

[13] Dobrzański LA, Domagała J, Tański T, Klimpel A, Janicki D (2008) Laser surface treatment of magnesium alloy with WC and TiC powders using HPDL. Journal of Achievements in Materials and Manufacturing Engineering. 28/2: 179-186.

[14] Dobrzański L.A, Malara Sz, Tański T, Klimpel A, Janicki D (2009) Laser surface treatment of magnesium alloys with silicon carbide powder. Archives of Materials Science and Engineering. 35/1: 54-60.

[15] Dobrzański LA, Tański T, Domagała J, Malara Sz, Klimpel A (2011) Laser surface treatment of Mg-Al-Zn alloys. Journal for Theory and Application in Mechanical Engineering. 53/1: 5-10.

[16] Dobrzańska-Danikiewicz AD (2010) Foresight methods for technology validation, roadmapping and development in the surface engineering area. Archives of Materials Science Engineering. 44/2: 69-86.

[17] Dobrzańska-Danikiewicz AD, Computer Aided Foresight Integrated Research Methodology in Surface Engineering Area, work in progress.

[18] Dobrzańska-Danikiewicz AD, Tański T, Malara Sz, Domagała-Dubiel J (2010) Assessment of strategic development perspectives of laser treatment of casting magnesium alloys. Archives of Materials Science Engineering. 45/1: 5-39.
[19] Dobrzańska-Danikiewicz AD, Lukaszkowicz K (2010) Technology validation of coatings deposition onto the brass substrate.Archives of Materials Science Engineering. 46/1: 5-38.
[20] Dobrzańska-Danikiewicz AD, Jonda E, Labisz K (2010) Foresight methods application for evaluating laser treatment of hot-work steels. Journal of Achievements in Materials and Manufacturing Engineering. 43/2: 750-773.
[21] Dobrzańska-Danikiewicz AD, Hajduczek E, Polok-Rubiniec M, Przybył M, Adamaszek K (2011) Evaluation of selected steel thermochemical treatment technology using foresight methods. Journal of Achievements in Materials and Manufacturing Engineering. 46/2: 115-146.
[22] Dobrzańska-Danikiewicz AD. Gołombek K, Pakuła D, Mikuła J, Staszuk M, Żukowska L (2011) Long-term development directions of PVD/CVD coatings deposited onto sintered tool materials. Journal of Achievements in Materials and Manufacturing Engineering. 49/2: 69-96.
[23] Dobrzańska-Danikiewicz AD, Drygała A (2011) Strategic development perspectives of laser processing on polycrystalline silicon surface. Archives of Materials Science Engineering. 50/1: 5-20.
[24] Dobrzańska-Danikiewicz AD, Kloc-Ptaszna A, Dołżańska B (2011) Determination of tool gradient materials value according to foresight methodology. Journal of Achievements in Materials and Manufacturing Engineering. 50/2: 69-96.
[25] Dobrzańska-Danikiewicz A (2010) E-foresight of materials surface engineering. Archives of Materials Science Engineering. 44/1: 43-50.
[26] The Boston Consulting Group, The Product Portfolio, Perspectives 66 (1970).
[27] Gerdsri N, Vatananan RS, DansamasatidS. (2009) Dealing with the dynamics of technology roadmapping implementation: A case study. Technical Forecasting & Social Change. 76: 50-60.
[28] Y. Yasunaga, M. Watanabe, M. Korenaga, Application of technology roadmaps to govermental innovation Policy for promoting technology convergence, Technical Forecasting & Social Change 76 (2009) 61-79.
[29] Phaal R, Muller G (2009) An architectual framework for roadmapping: Towards visual strategy. Technological Forecasting & Social Change. 76: 39-49.
[30] Dobrzańska-Danikiewicz AD, Jonda E, Trzaska J, Jagiełło A, Labisz K (2011) Neural network aided future events scenarios presented on the example of laser surface treatment. Journal of Achievements in Materials and Manufacturing Engineering. in press.

Section 3

Orthopaedic Applications – Magnesium Alloys

Rare Earth Metals as Alloying Components in Magnesium Implants for Orthopaedic Applications

Nina Angrisani, Jan-Marten Seitz,
Andrea Meyer-Lindenberg and Janin Reifenrath

Additional information is available at the end of the chapter

http://dx.doi.org/10.5772/48335

1. Introduction

During the last decade magnesium and magnesium alloys (MA) were the centre of a large number of studies especially in Europe [1-3] and Asia [4-6]. The main focus of these studies was to evaluate the use of magnesium and MA as basic material for clinical applications.

For the treatment of bone fractures, orthopaedic implants made of surgical steel or titanium are used when weight bearing bones are affected. The major disadvantage of these materials is that they need to be removed in a second surgery due to implant loosening or intolerance after longer implantation times resulting in higher costs and stress for the patient. Therefore resorbable implant materials are needed which complementarily provide sufficient stability for weight bearing applications.

Magnesium is a light metal which is known to corrode in aqueous solution. Its density is 1.74 g/cm^3 at room temperature and therewith 1.6 fold resp. 4.5 fold lower than aluminium or steel [7]. With prospect of orthopaedic use, its advantages are its appropriate compressive and tensile strengths as well as its Young's Modulus (41-45 GPa), which is considerably closer to cortical bone compared to other metallic implant materials [7-9]. The corrosion rate ranges between aluminium and unalloyed steel [10].

Magnesium as a mineral occurs naturally in the body and is eliminated through the kidneys [11,12]. Due to their high excretion ability hypermagnesaemia is rare [13]. Severe symptoms like arrhythmia, feeling of faintness up to paralyses and/or cardiac or respiratory arrest are only seen in the course of therapeutical intravenous application [11,14].

Therefore, magnesium and MA are intensively investigated to develop a basic material for the production of degradable osteosynthesis implants e.g. plates, screws or intramedullary nails.

Alloying with other elements such as lithium, aluminium, zinc or rare earth metals aims to adjust the corrosion resistance and mechanical properties of magnesium [9,15-18]. In engineering applications **Aluminium** (Al) as another light metal is often used as alloying component in MA for its beneficial effect of strength, hardness and castability improvement [19]. For an optimum balance between strength and ductility the authors claim a content of 6 wt%. Contents between 1-9 wt% are classified as corrosion protective, whereof the higher the content the better the protection [8]. Many groups investigated MA containing Al [20-27], despite its questionable biocompatibility [28].High concentrations are considered to be neurotoxic and implicated in pathologies such as dementia, senile dementia and Alzheimer's Disease [29]. However according to reference [30] the uptake of even high amounts of Al results in physiological neutral behaviour and exceptional low quantities are absorbed. Formerly, different groups reported on high corrosion rates of Mg-Al-alloys [20-22], but recently a decrease in corrosion rate could be shown in comparison to pure Mg or other MA, especially when further elements e.g. zinc [26,27] or Rare Earth Metals [23,31] are added.

Besides aluminium, **Lithium** (Li) could be added to increase the ductility and the corrosion resistance with a simultaneous decrease in strength [8,19,32]. However, the engineering application is limited [19]. In medicine, Li is therapeutically used for the treatment of manic-depressive disorders. However, a correlation with teratogenicity, nephrotoxicity and mania is discussed [33].

Zinc is often used in combination with aluminium to improve strength at room temperature [19]. In combination with zirconium or rare earths it is used to produce MA hardenable by precipitations and with superior strength. Regarding the corrosion behaviour, zinc lowers the corrosive effect of iron and nickel impurities [19]. In the body zinc belongs to the essential trace elements and is excreted mainly via the faeces. It plays an important role in numerous processes e.g. protein synthesis, nucleic acids synthesis, carbohydrate and lipid metabolism [34]. Zinc deficiency is associated with an increase in bone mass and osteoblast DNA synthesis [35,36].

The main reason for the addition of **Zirconium** to MA is the purpose of grain refinement. Although it is a powerful grain refiner it cannot be used in Al-containing MAs because it is removed from solid solution due to the formation of stable compounds [19]. Such compounds are also formed with elements like manganese, iron, silicon, carbon, nitrogen, oxygen or hydrogen when they are present within the melt [19]. Hence, the amount of soluble zirconium is the important factor rather than the total amount. Zirconia implants have excellent resistance to corrosion and wear, good biocompatibility and high bending strength and fracture toughness [37].

The usual addition of **Rare Earths** (RE) in engineering applications is performed as mischmetal or didymium, whereof the mischmetal contains 50 wt% cerium and the rest principally neodymium and lanthanum [19]. REs aim to increase the strength of MA and to decrease weld cracking and porosity during the casting procedure [19]. Regarding the development of a resorbable implant material it is important, that the addition of RE can achieve an increase of the corrosion resistance [23,38-40]. However, this also depends on the other alloying elements. In reference [41] was reported on a decrease of the corrosion resistance in vitro after combining Al and Neodymium with Mg as main alloying component.

There are controversial reports about the effect and toxicity of REs. Especially high concentrations are considered to have toxic effects [42-45].On the other hand a bone-protective effect and an increase in bone density could be shown after a six months feeding trial [46]. Low amounts of REs in MA were appraised as well tolerable in reference [28]. In reference [45] the short-term effect of REs in vitro was evaluated. The authors assessed the responses of different cell lines after the addition of the single RE elements and found differences between light (Lanthanum (La), Cerium (Ce), Praseodymium (Pr)) and medium to heavy (Neodymium (Nd), Europium (Eu), Gadolinium (Ga) and Dysprosium (Dy)) RE elements: light RE elements showed toxic effects at lower concentrations [45]. They concluded that La and Ce should be used only when absolutely necessary. Most of the contemporarily published in vivo examinations used MA with low amounts of REs, which were added as mischmetal [2,9,40,47-50]. This mischmetal can actually differ depending on the time and/or date of purchase. For example, [45] determined the RE mixture of WE43 to be mainly Nd, Gd and Dy whereas in the most commercially available RE composition metals Ce, Nd and La form the major fraction [44,51]. In *in vivo* examinations, LAE442, a MA with 4 wt% lithium, 4 wt% aluminium and 2 wt% REs showed generally good biocompatibility with slow and homogenous degradation properties [1,2,23,47]. Nevertheless, in reference [3] was pointed out, that from a medical point of view the addition of such a content-varying mixture has to be seen very critically since reproducibility is one of the main requirements for medical devices. Hence, they examined the in vivo degradation behaviour of this repeatedly used MA LAE442 in comparison to LACe442 which replaced the RE mixture by the single RE element Cerium. The outcome of this in vivo study supported the simultaneously performed examinations described in reference [45] as this replacement led to a severe increase of the degradation rate with subsequent tissue reactions. Therefore the authors concluded that Ce could not supersede the RE composition. So far unknown regulative effects between the different RE elements seem to exist.

But despite the good results of LAE442, the effort to replace the mixture by a single element is still reasonable to achieve a most accurate implant device.

Since in reference [45] Nd was classified as suitable, referring to LAE442, LANd442 was developed as well as Nd2 and their in vitro corrosion behaviour was assessed [41]. They showed that the corrosion rate for Nd2 was lower than for LANd442. However, the application of a MgF_2-coating lowered the corrosion rate of LANd442. After these positive results, LANd442 was introduced into in vivo experiments. Parts of this study will be included in the next subchapter.

Some groups reported on the fact that in vitro and in vivo degradation properties could differ considerably [6,24]. Thus, besides profound in vitro tests like bending tests, corrosion tests or microstructure characterisation to identify and adjust the material properties, the in vivo degradation behaviour and biocompatibility has to be investigated thoroughly.

Besides other animal models e.g. rats and sheep, the rabbit is a well established laboratory animal in orthopaedic research [52-54].

2. *In vivo* experiments

To investigate the clinical applicability of different RE containing MA the rabbit (female New Zealand White rabbits, body weight > 3 kg) was chosen as animal model. Partially, the results of the following data have already been published or submitted [55-57]. Extruded, cylindrical pins (length 25 mm, diameter 2.5 mm) were produced and washed in acetone for eliminating fabrication residues. Sterilization was carried out by gamma irradiation. Ten implants of three different RE containing MA were produced: LAE442, LANd442 and ZEK100. These cylinders were randomly implanted into the middle third of the tibial medullary cavity (Fig 1), one in each hind leg. The follow up period was six months. Clinical examinations were performed regularly to assess the clinical tolerance of the implants.

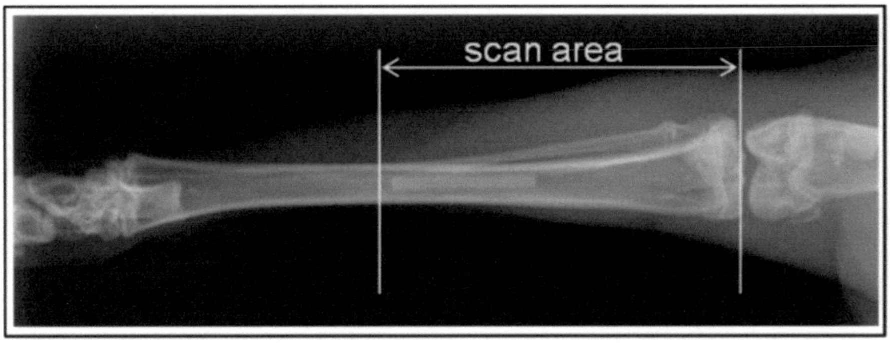

Figure 1. X-ray depiction of an implanted MA-pin with marked μ-CT scan area.

μ-Computed tomography is a non-destructive analysing method to assess changes within the structure of either engineering components or medical implants [2,40,58,59]. It can also be used to evaluate the reactions of the surrounding tissue to implanted orthopaedic devices after in vivo examinations [57,60,61]. However, the laboratory animals have to be sacrificed for these investigations. To evaluate the proceeding degradation and to perform a proper initial-to-end-value comparison in vivo μ-computed tomography scans have been introduced recently [50,55,57,60,62]. Further they allow for a reduction of laboratory animals as the results after different implantation periods could be gained from the same animal.

For the presented study an XtremeCT (Fa. Scanco Medical, Zurich, Switzerland) was used. The animals were scanned in general anaesthesia and in supine position (Fig.2). The scan was performed from the knee joint space up to approx. 5 mm beneath the implant (Fig.1) with a resolution of 41 μm, 1000 projections at 0-180° and an integration time of 100 ms. The electron energy used was 60 kVp and the intensity was 900 μA. In the first eight weeks the rabbits were scanned biweekly, afterwards every four weeks.

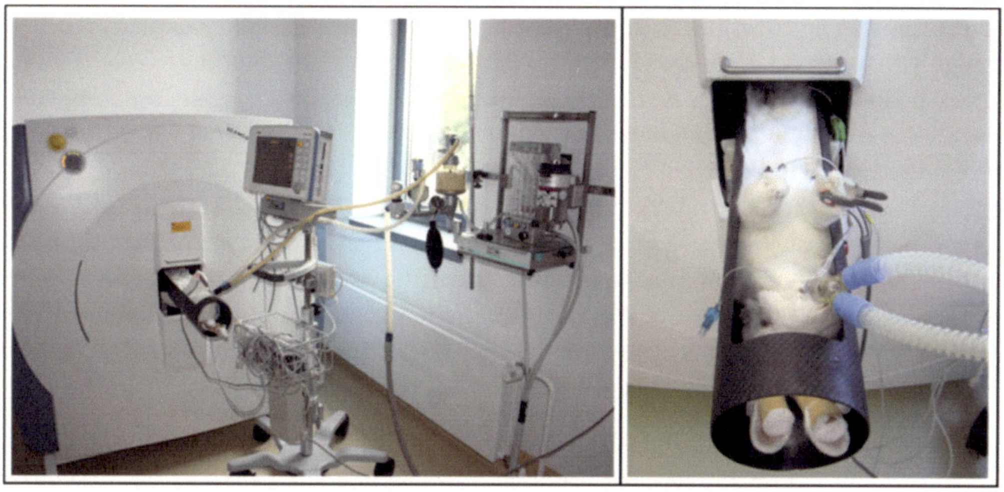

Figure 2. μ-computed tomography of the rabbit tibia under general anaesthesia.

The μ-computed tomography evaluation **included three parts**: First, the implant itself was assessed (changes in structure and volume of the pins as well as their corrosion morphology). For each implant material a specific threshold was determined which represented the pin most accurately (LAE42 and LANd442: 138, ZEK100: 127).The implants were subsequently manually outlined and measured by means of the software μCT evaluation program V6.1 (XtremeCT, Fa. Scanco Medical, Zurich, Switzerland). To further quantify the corrosion rate and the corrosion morphology, the direct 3D thickness of the volumes of interest (VOIs) was calculated. Therefore, the structure was filled with overlapping spheres of maximal diameter. The diameter of the spheres at tach location denotes the local thickness. The average thickness was determined by averaging over the whole structure resulting in histograms of bin sizes with an average 3D thickness and a standard deviation for each implant. A low average bin size with a low standard deviation indicates a high degree of uniform corrosion. A high standard deviation of the histogram is caused by an irregular shape of the remaining implant and therefore it is an indicator for the extent of pitting corrosion [55].

Second the gas which emerged during the degradation process of magnesium implants was assessed. The corrosion mechanism of pure Mg and MA consists of two electrochemical parts: the anodic partial reaction forms Mg^{2+} and $2e^-$, whereas the cathodic partial reaction evolves hydrogen and $2OH^-$ from the reaction of water with the $2e^-$ [10,63,64]. In vitro corrosion tests of MA which quantify the amount of emerging gas utilize this mechanism [41]. Many in vivo studies reported on the emergence of gas during the course of MA-implant degradation either as diffuse accumulation or palpable and non-palpable gas bubbles underneath the skin [23,55,58,65]. It is a continuous discussion, if these bubbles actually contain hydrogen. Hence, as a supplementary investigation a gas-tight syringe was used to gather the emerged gas (approx. 0.5 ml) out of a large subcutaneous gas bubble. It was sent to the Institute of Organic Chemistry, Technical University Braunschweig, and analyzed.

However, it was not possible to verify pure hydrogen. The most likely explanation is that the highly volatile gas undergoes rapid exchange with the surrounding tissue. From the authors' point of view there is no need to doubt that the degrading magnesium implants are the source of the gas independent on the actual composition in the bubbles as faster degrading alloys subjectively generate higher amounts of gas. This leads to the matter of quantifying the gas volume. So far, no method has been described to report on the quantity of gas emerged particularly over the course of time. So within this second part of the µCT evaluation a method for quantifying the gas volume using the XtremeCT was established. Within the 2D-slices of the µCT-scans of the LAE442 group, the occurring gas was manually outlined and measured by means of the software µCT evaluation program V6.1. The threshold of the grey values was determined to be between -1000 and 25.

With proceeding degradation the corrosion products influence the surrounding tissue. The smaller the impact of the implant the better is its biocompatibility. On the one hand it could be generally said, that an implant is biocompatible when its functionality is achieved without inducing a foreign body reaction [66]. On the other hand, according to the Conference 1984 of the European Society for Biomaterials, biocompatibility is the ability of a material to fulfil its purpose for a specific application with an appropriate host response [67]. This definition includes the fact that every inserted implant actually could/will influence the surrounding tissue in one or another way and emphasize on the adequacy. The reactions which are described are either foreign body reactions ([3,57,60,65] or structural changes of the bone [2,57,61]. To assess cellular reactions histological examinations have to be carried out. However, µCT is a well-established tool to evaluate structural changes of the bone. Thus, as a third part of the µCT evaluation the impact of the degrading implant on the adjacent bone was evaluated by a quantitative determination of the bone density (in mg HA/cm3), the bone volume (in mm^3/slice) and the bone porosity (in percent). The bone volume which was included into the evaluation was defined by choosing those slices in which also the implant was seen. Hence, the bone directly adjacent to the implant was manually outlined. The threshold value for the subsequent evaluations was determined to be 160 and the same software was used as for the implant and gas evaluation. Due to internal processing, the latest investigation time for ZEK100 was week 20.

2.1. Pin degradation

The density is given in the unit mg HA/mm^3 (milligram hydroxyapatite per cubic millimetre) which is the unit the XtremeCT gives for the density of mineralized tissue such as bone. Therefore, the indicated density values do not correspond to common used alloy density values but allow for a comparison of the three alloys investigated among each other. Their density differed from the beginning of the implantation period. LAE442 density was higher than LANd442. ZEK100 showed the lowest density (Fig. 3). Besides the varying alloying components different grain sizes of the alloy could influence the density [68] and therefore cause the detected differences. During the course of degradation LAE442 showed only a slight decrease in density with a very low standard deviation. Also the density of

LANd442 implants diminished slightly, however the standard deviation was obviously higher implying a more inhomogeneous procedure. ZEK100 implants showed the highest loss in density.

The initial volume of all alloys ranged in similar values. According to the density, LAE442 implants showed only a minor decrease over the implantation period and demonstrated again a low standard deviation. The changes of volume in LANd442 implants also matched the results of the density. A slight decrease could be found with a higher standard deviation in comparison to LAE442. ZEK100 pins demonstrated an obvious loss of volume particularly from the 12th week on with an exceptional high standard deviation in the later scan weeks (Fig. 3).

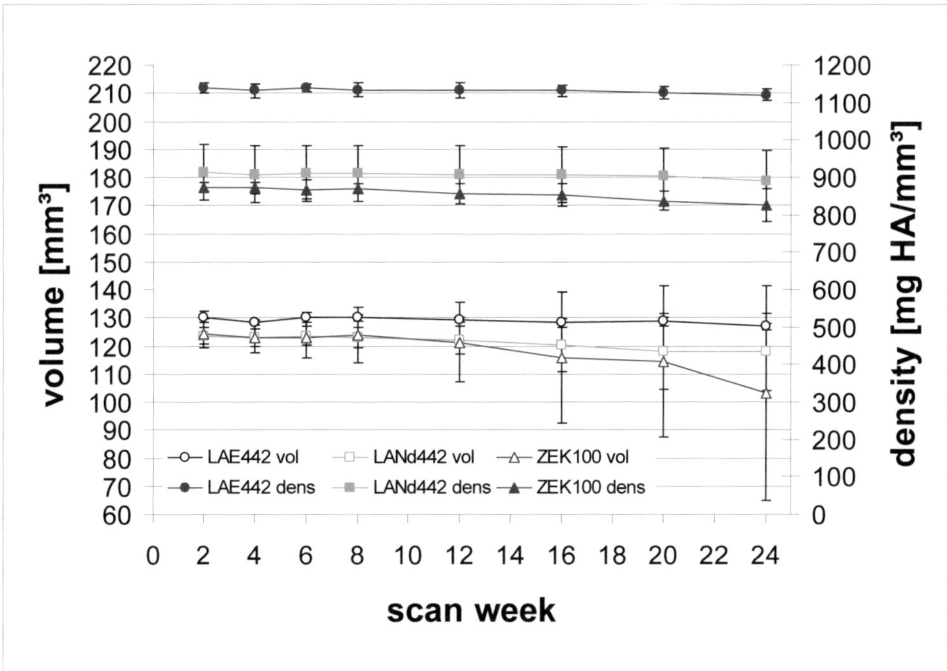

Figure 3. Volume and density changes of LAE442, LANd442 and ZEK100 implants over the implantation duration of 24 weeks.

These results indicate a slow and uniform degradation of LAE442 implants. LANd442 pins also degraded slowly but less uniformly. ZEK100 showed an equally slow degradation within the first weeks of implantation. This process accelerated distinctly resulting in inhomogeneous pin geometries.

The examination of the true 3D-thickness confirmed these findings (Fig. 4). Corresponding to the volume and density changes, the average diameter of the spheres and thus the true 3D-thickness of LAE442 implants underlay only minor changes in the course of degradation. The low and uniform variance of diameter is a sign for a very homogeneous degradation. LANd442 implants showed a slight decrease of the true 3D-thickness. The variance of diameter increased in the course of implantation moderately. Taken together both results it could be said that LANd442 pins degraded slowly but faster and more inhomogeneous than LAE442. ZEK100 displayed the most obvious changes. From the 8th week on the average diameter of the spheres decreased continuously while the variance of diameters showed a profound increase. Consequently, ZEK100 implants degraded fast and irregularly.

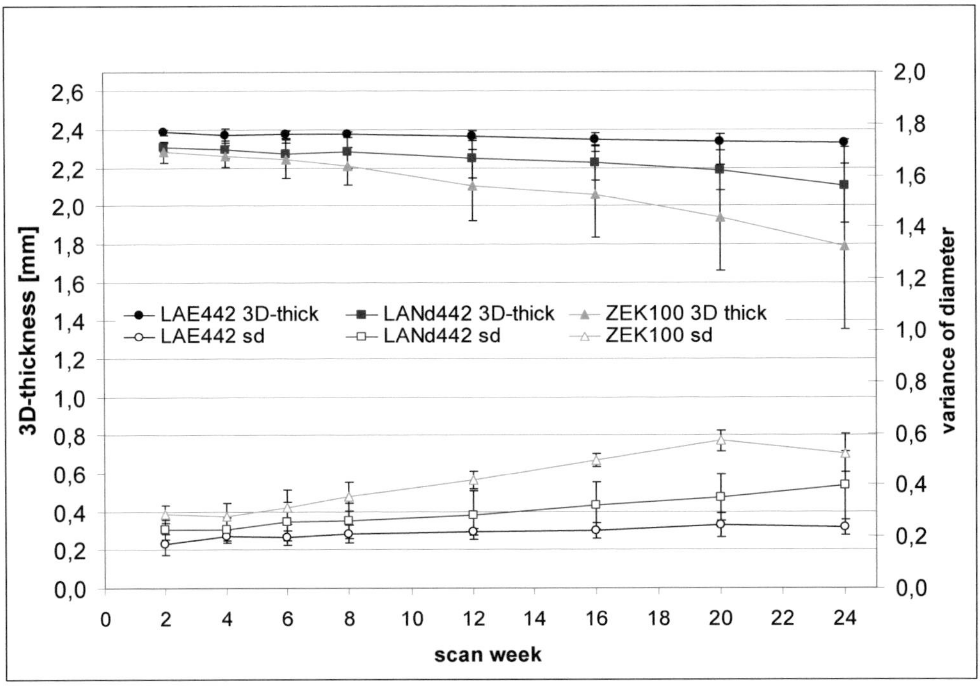

Figure 4. True 3D-thickness and variance of diameter as indicator for homogeneous or heterogeneous degradation of the implanted LAE442, LANd442 and ZEK100 pins, respectively.

The colour mapping of the degraded implants after six months implantation duration in comparison to an undegraded implant visualized the differences of the pin geometry (Fig. 5).

2.2. Gas volume

In outlining and evaluating the occurring gas within the marrow cavity it was possible to quantify the gas volume during the course of degradation.

It is noteworthy that a distinct proceeding decrease in volume was found until week 12 followed by a continuous increase. This could be explained by the fact that a certain amount of gas was brought into the marrow cavity due to the surgical procedure. This gas volume is reabsorbed by the organism in the subsequent time. Corresponding to the beginning degradation of the implant, which is represented by the volume and density changes (Fig. 6 and 7), the amount of gas which is emerged exceeded the absorption capacity of the organism resulting in the increase of gas volume.

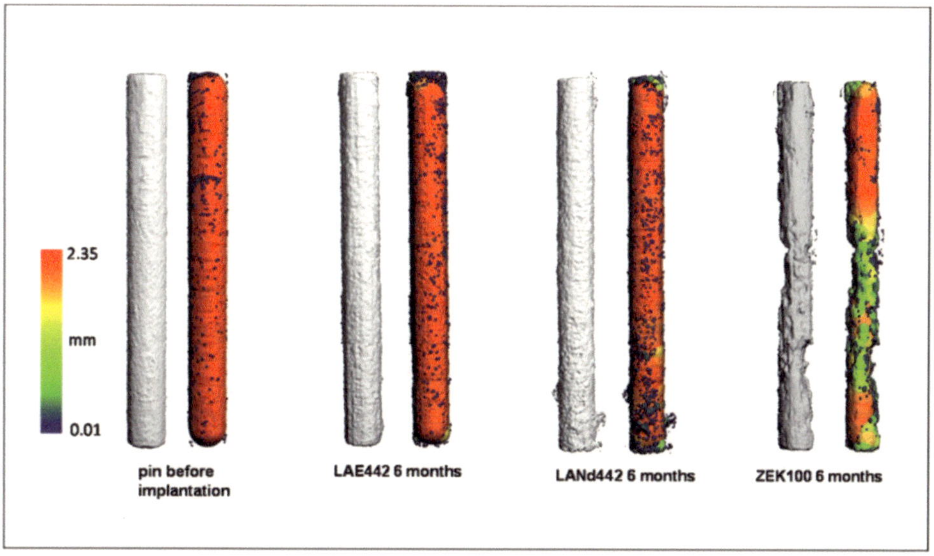

Figure 5. 3D-depiction and colour mapping of a MA-cylinder before implantation and of MA-cylinders (alloys: LAE442, LANd442 and ZEK100) after six months implantation period in the rabbit tibia.

Fig. 8 shows a 3D-evaluation of the bone (transparent) with implanted MA-cylinder (blue) and surrounding gas (brown).

For further examinations, this method can be used to compare the gas volume with possible occurring changes of the surrounding tissue.

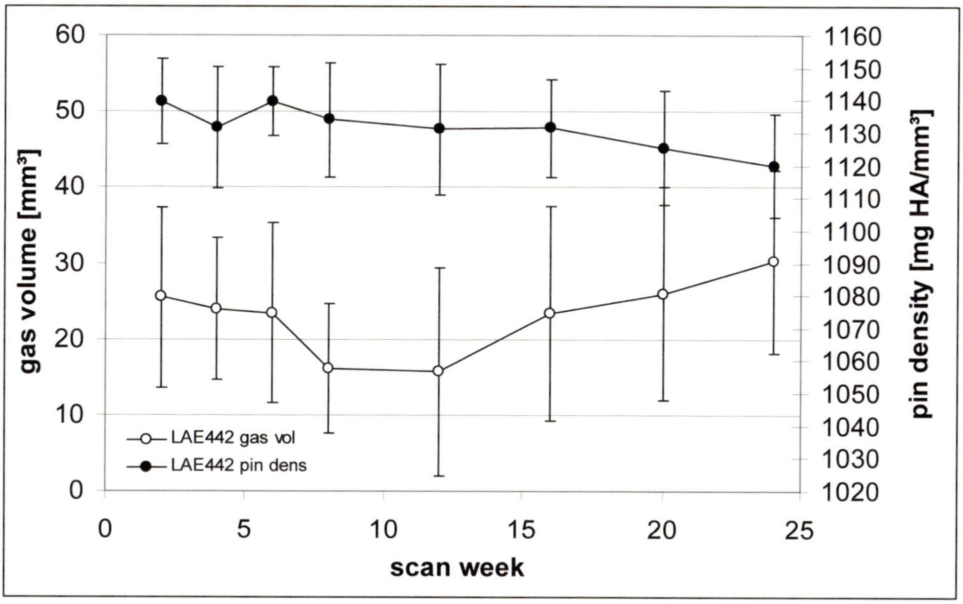

Figure 6. Quantity of gas volume and pin density of LAE442 cylinders over six months implantation duration measured by μ-computed tomography (XtremeCT, Scanco medical).

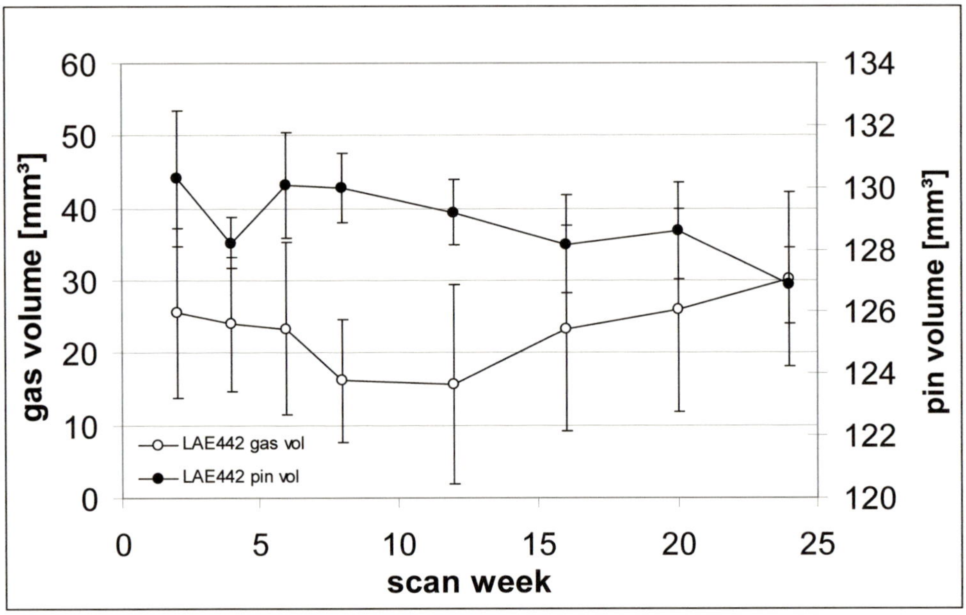

Figure 7. Quantity of gas volume and pin volume of LAE442 cylinders over six months implantation duration measured by μ-computed tomography (XtremeCT, Scanco medical).

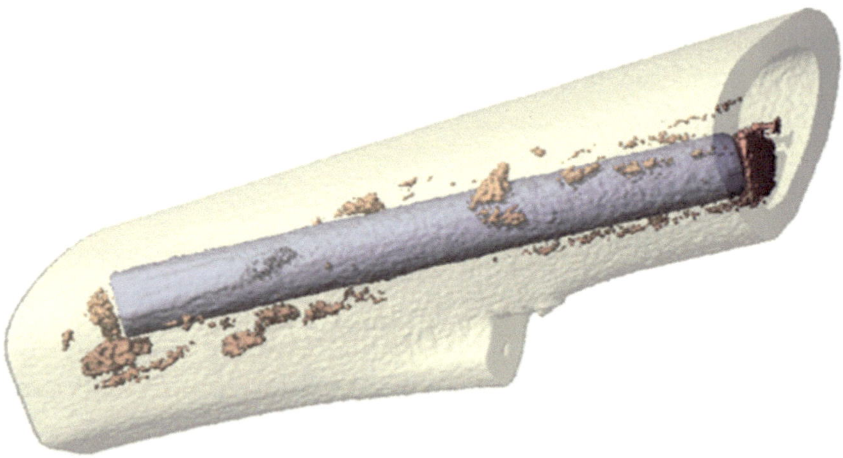

Figure 8. 3D-depiction of an implanted MA-cylinder (blue) within the tibia diaphysis (yellow-transparent) surrounded by gas (brown-red).

2.3. Changes in bone structure

The bone adjacent to all three types of implanted MA pins lost density during the course of degradation (Fig. 9). For LAE442 implants the decrease in density was more pronounced in the first weeks of implantation and slowed down in the following period. From week 12 on only a negligible further decrease could be seen. LANd442 also showed a moderate decrease in bone density up to week 12 followed by an increase in the next four weeks and subsequent steady state until week 24. Due to the lower number of investigations the density course of ZEK100 implants appeared to be different. They induced a proceeding reduction of the bone density over the total investigation period. However, a higher number of investigations could reveal a similar pattern as for LAE442 and LANd442.

The bone volume (specified per slice) increased over the investigated time period.

No distinct differences could be found for the different MA cylinders (Fig. 10).

In contrast to the aforementioned results the changes in porosity of the bone showed no regular pattern (Fig. 11). The bone porosity adjacent to LAE442 implants first increased up to week 4, followed by a decrease up to week 12. Afterwards the bone porosity again increased up to the end of the investigation period but slower than in the beginning. LANd442 implants showed only minor changes within the first 12 weeks. After that the porosity increased first moderately up to week 16 and from then on intensely till the end of investigation. However this increase in the LANd442 mean value is particularly caused by one implant which degraded severely faster than all other implants of the same group without any obvious explanation. The high standard deviation for the scans in week 16 and 24 illustrated this fact. If this cylinder would be excluded from the evaluation the mean porosity of the LANd442 pins would steadily decrease from week 8 on up to week 24. After the implantation of ZEK100 cylinders a distinct increase in bone porosity could be seen in the first eight weeks followed by a moderate decrease up to week 20.

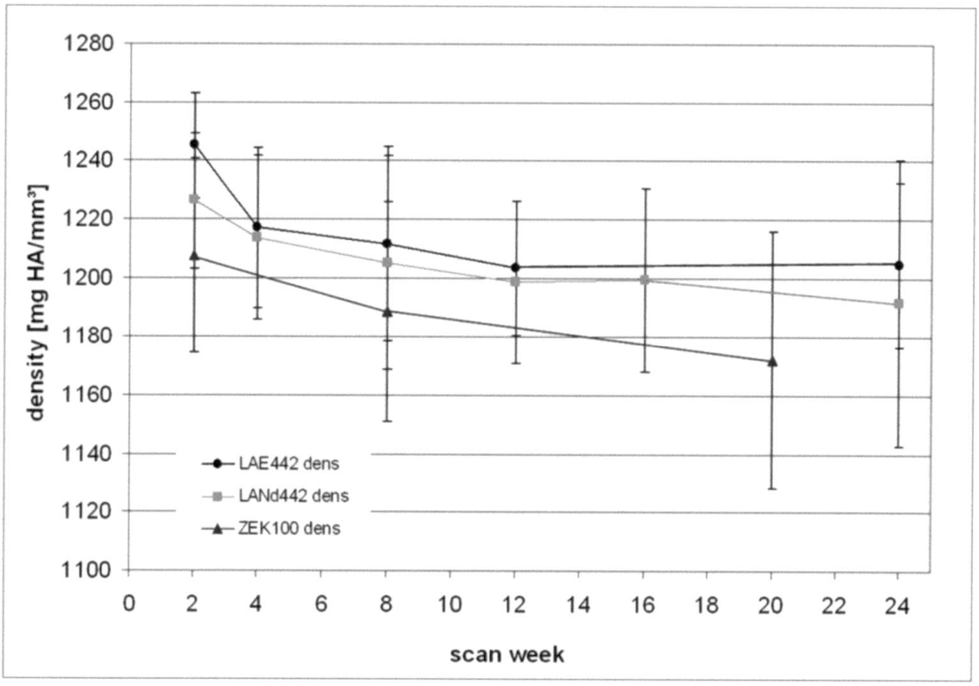

Figure 9. Bone density adjacent to implanted LAE442-, LANd442- and ZEK100-cylinders in the course of implantation over up to 24 weeks.

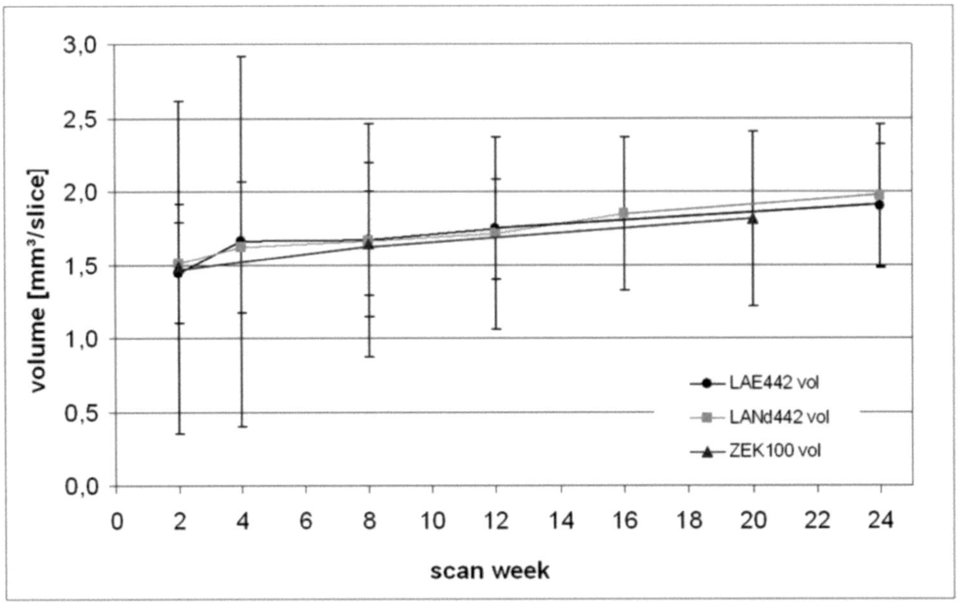

Figure 10. Bone volume/slice adjacent to implanted LAE442-, LANd442- and ZEK100-cylinders in the course of implantation over up to 24 weeks.

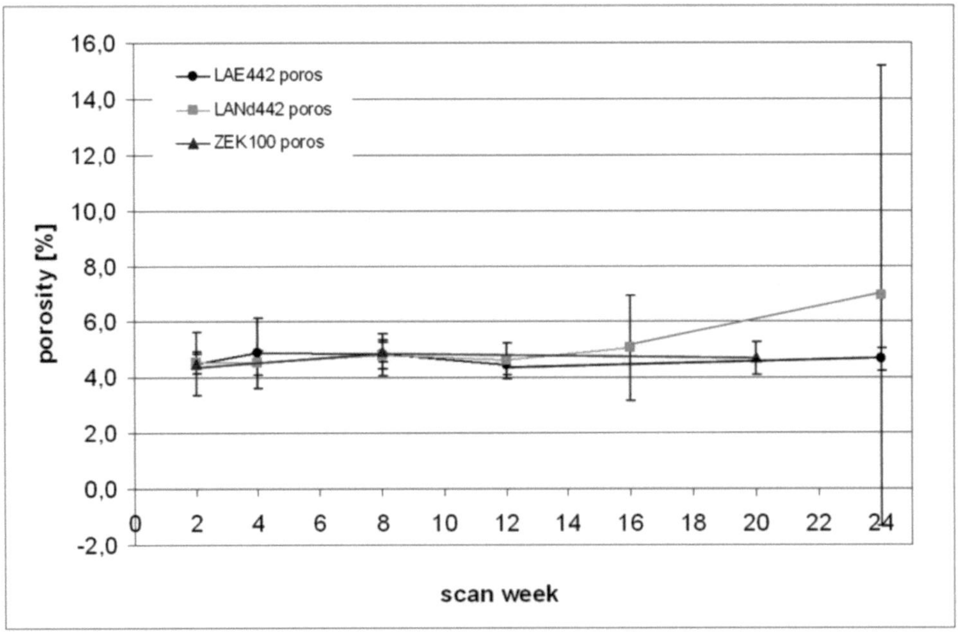

Figure 11. Bone porosity adjacent to implanted LAE442-, LANd442- and ZEK100-cylinders in the course of implantation over up to 24 weeks.

Taken into account all gathered results the μ-computed tomography evaluation of the bone structure illustrated the bone remodelling processes well. The increase in bone volume indicated endosteal and periosteal new bone growth. The simultaneous decrease in bone density can be explained by the fact that this newly formed bone is not as dense as the mature bone. The varying porosity depends on two different factors. On the one hand the state of the bone remodelling process influences this value. The newly formed bone is not as much structured as the mature bone and therefore shows a higher porosity. Since the bone of LAE442 implants showed a higher increase in bone volume accompanied by the most distinct decrease in density particularly in the first weeks of implantation the ascending porosity at this time could be explained. On the other hand the two-dimensional evaluation which was not included here showed that the faster degrading ZEK100 implants induced more bone cavities than the slower degrading alloys [65]. This is probably the cause for the initial increase in porosity. To which extend the decreasing density influences the computation of the porosity remains to be shown as the further degrading ZEK100 cylinders should continuously increase the porosity instead of the determined however slight decrease.

3. Conclusions

After considering both the accessible literature and the presented results, REs seem useful and maybe even necessary alloying components in MA basis material for the production of orthopaedic implants.

It could be shown that the addition of REs lead to mostly biocompatible, degradable implants with however different degradation characteristics.

LAE442 proved to be the slowest degrading alloy with the lowest influence on the surrounding tissue. The replacement of the RE composition metal by the single element Nd did not result in improvement of the biocompatibility nor of the degradation behaviour. Contrary, these LANd442 implants showed a less regular corrosion process than LAE442 cylinders. Therefore LAE442 implants should be favoured over LANd442. Therewith, two alloys which aimed to replace the RE composition by a single element (LANd442, LACe442 [3]) failed in exceeding the good degradation behaviour and biocompatibility of LAE442.

ZEK100 as a completely different approach to develop an alloy for orthopaedic implant production clearly showed inferior results by degrading inhomogeneously and causing significant structural changes in the adjacent bone and tissue [55].

In principal, a slow degrading MA should be developed as a faster degradation is correlated with a reduced clinical tolerance and an increased impact on the adjacent bone: the implant in the right leg of one LANd442 rabbit degraded significantly faster than the other implants. The animal showed a moderate to severe lameness of the affected leg and an obvious increase in bone porosity. Also in reference[3] was reported on the poor biocompatibility of LACer442 which degraded very fast.

Altogether, the perfect composition of a degradable MA implant material has not been developed yet. Regarding the biocompatibility, Li-Al-RE-containing MA turned out to be very promising. However, to satisfy the high standards for the production of medical devices efforts to replace the RE composition metal by one or even a couple of the single RE elements should not be abandoned with further focus on a slow and homogenous degradation behaviour.

Author details

Nina Angrisani and Janin Reifenrath
Clinic for Small Animals, University of Veterinary Medicine, Hannover, Germany

Jan-Marten Seitz
Institute of Materials Science, Leibniz University Hannover, Garbsen, Germany

Andrea Meyer-Lindenberg
Clinic for Small Animal Surgery and Reproduction, Centre of Clinical Veterinary Medicine, Faculty of Veterinary Medicine, Ludwig-Maximilians-University Munich, Munich, Germany

Acknowledgement

The authors thank Dr. Ulli Papke from the Institute of Organic Chemistry (TU Braunschweig) for analyzing the gas out of the gas-tight syringe.

The presented results are part of the collaborative research centre (CRC599, Medical University of Hannover, University of Veterinary Medicine Hannover and University of Hannover), which is sponsored by the German Research Foundation (DFG).

4. References

[1] Witte F, Ulrich H, Rudert M, Willbold E (2007) Biodegradable magnesium scaffolds: Part 1: appropriate inflammatory responseJ Biomed Mater Res A 81: 748–756.

[2] Thomann M, Krause C, Bormann D, Höh N von der, Windhagen H, Meyer-Lindenberg A (2009) Comparison of the resorbable magnesium alloys LAE442 und MgCa0.8 concerning their mechanical properties, their progress of degradation and the bone-implant-contact after 12 months implantation duration in a rabbit modelMat Sci Eng Tech 40: 82–87.

[3] Reifenrath J, Krause A, Bormann D, Rechenberg B von, Windhagen H, Meyer-Lindenberg A (2010) Profound differences in the in-vivo-degradation and biocompatibility of two very similar rare-earth containing Mg-alloys in a rabbit model Mat.-wiss. u. Werkstofftech. 41: 1054–1061.

[4] Xu L, Yu G, Zhang E, Pan F, Yang K (2007) In vivo corrosion behavior of Mg-Mn-Zn alloy for bone implant application J Biomed Mater Res A 83: 703–711.

[5] Li Z, Gu X, Lou S, Zheng Y (2008) The development of binary Mg-Ca alloys for use as biodegradable materials within bone Biomaterials 29: 1329–1344.

[6] Zhang S, Zhang X, Zhao C, Li J, Song Y, Xie C, Tao H, Zhang Y, He Y, Jiang Y, Bian Y (2010) Research on an Mg-Zn alloy as a degradable biomaterial Acta Biomater 6: 626–640.
[7] Catrin Kammer (2000) Magnesium-Taschenbuch. Düsseldorf: Aluminium-Verlag.
[8] Volker Kaese (2002) Beitrag zum korrosionsschützenden Legieren von Magnesiumwerkstoffen. Düsseldorf: VDI.
[9] Staiger MP, Pietak AM, Huadmai J, Dias G (2006) Magnesium and its alloys as orthopedic biomaterials: a review Biomaterials 27: 1728–1734.
[10] Song G, Atrens A (1999) Corrosion mechanisms of magnesium alloys Adv Eng Mater 1: 11–33.
[11] Hermans C, Lefebvre C, Devogelaer JP, Lambert M (1996) Hypocalcaemia and chronic alcohol intoxication: transient hypoparathyroidism secondary to magnesium deficiency Clin Rheumatol 15: 193–196.
[12] Hartwig A (2001) Role of magnesium in genomic stability Mutation research 475: 113–121.
[13] Tinsley R Harrison, Anthony S Fauci (1998) Harrison's principles of internal medicine. New York: McGraw-Hill Health Professions Div.
[14] Knochel J (1998) Disorders of magnesium metabolism. In: Harrison TR, Fauci AS, editors. Harrison's principles of internal medicine. New York: McGraw-Hill Health Professions Div. pp. 2263–2266.
[15] Drynda A, Deinet N, Braun N, Peuster M (2009) Rare earth metals used in biodegradable magnesium-based stents do not interfere with proliferation of smooth muscle cells but do induce the upregulation of inflammatory genes J Biomed Mater Res A 91: 360–369.
[16] Gu X, Zheng Y, Cheng Y, Zhong S, Xi T (2009) In vitro corrosion and biocompatibility of binary magnesium alloys Biomaterials 30: 484–498.
[17] Zberg B, Uggowitzer PJ, Löffler JF (2009) MgZnCa glasses without clinically observable hydrogen evolution for biodegradable implants Nat Mater 8: 887–891.
[18] Yang Y, Li X (2010) Influence of neodymium on high cycle fatigue behavior of die cast AZ91D magnesium alloy Journal of Rare Earths 28: 456–460.
[19] Polmear IJ (1999) Grades and Alloys In: Avedesian MM; Baker H, editors. Magnesium and Magnesium Alloys. Materials Park, OH: ASM International. pp. 12–23.
[20] Verbrugge J (1934) Le matériel métallique résorbable en chirurgie osseuse La Press Med 23: 460–465.
[21] McBride ED (1938) Absorbable metal in bone surgery J Am Med Assoc 111: 2464–2467.
[22] Znamenskii MS (1945) Metallic osteosynthesis by means of an apparatus made of resorbing metal Khirurgiya 12: 60–63.
[23] Elinor Switzer (2005) Resorbierbares metallisches Osteosynthesematerial. Available: http://elib.tiho-hannover.de/dissertations/switzere_ss05.pdf
[24] Witte F, Fischer J, Nellesen J, Crostack HA, Kaese V, Pisch A, Beckmann F, Windhagen H (2006) In vitro and in vivo corrosion measurements of magnesium alloys Biomaterials 27: 1013–1018.

[25] Witte F, Reifenrath J, Müller PP, Crostack HA, Nellesen J, Bach F-W, Bormann D, Rudert M (2006) Cartilage repair on magnesium scaffolds used as a subchondral bone replacement Mat.-wiss. u. Werkstofftech. 37: 504–508.

[26] Huang J, Ren Y, Jiang Y, Zhang B, Yang K (2007) In vivo study of degradable magnesium and magnesium alloy as bone implant Front. Mater. Sci. China 1: 405–409.

[27] Pardo A, Merino MC, Coy AE, Arrabal R, Viejo F, Matykina E (2008) Corrosion behaviour of magnesium/aluminium alloys in 3.5 wt.% NaCl Corros Sci 50: 823–834.

[28] Song G (2007) Control of biodegradation of biocompatable magnesium alloys Corros Sci 49: 1696–1701.

[29] El-Rahman SS (2003) Neuropathology of aluminum toxicity in rats (glutamate and GABA impairment). Pharmacol Res 47: 189–194.

[30] Ungethüm M, Winkler-Gniewek W (1984) Toxikologie der Metalle und Biokompatibilität metallischer Implantatwerkstoffe Z Orthop Ihre Grenzgeb 122: 99–105.

[31] Krause C, Bormann D, Hassel T, Bach F-W, Windhagen H, Krause A, Hackenbroich Chr (2006) Mechanical properties of degradable magnesium implants in dependence of the implantation duration. In: Proceedings of the international symposium on magnesium technology in the global age: 329–343.

[32] Haferkamp H, Kaese VNM, Philipp K, Phan-Tan THB, Rohde R (2001) Untersuchungen an magnesiumbasierten Legierungen als neue Materialien in der Implantologie Mat.-wiss. u. Werkstofftech. 32: 116–120.

[33] McIntyre RS, Mancini DA, Parikh S, Kennedy SH (2001) Lithium revisited Can J Psychiatry 46: 322–327.

[34] Islam MS, Du Loots T (2007) Diabetes, metallothionein, and zinc interactions: a review Biofactors 29: 203–212.

[35] Litchfield TM, Ishikawa Y, Wu LN, Wuthier RE, Sauer GR (1998) Effect of metal ions on calcifying growth plate cartilage chondrocytes Calcif. Tissue Int. 62: 341–349.

[36] Lutz W, Burritt MF, Nixon DE, Kao PC, Kumar R (2000) Zinc increases the activity of vitamin D-dependent promoters in osteoblasts Biochem. Biophys. Res. Commun. 271: 1–7.

[37] Rocchietta I, Fontana F, Addis A, Schupbach P, Simion M (2009) Surface-modified zirconia implants: tissue response in rabbits Clin Oral Implants Res 20: 844–850.

[38] Wu G, Fan H, Zhai C, Zhou YP (2005) The effect of Ca and rare earth elements on the microstructure, mechanical properties and corrosion behaviour of AZ91D Mater Sci Eng A 408: 255–263.

[39] Kannan MB, Raman RKS (2008) In vitro degradation and mechanical integrity of calcium-containing magnesium alloys in modified-simulated body fluid Biomaterials 29: 2306–2314.

[40] Krause A, Höh N von der, Bormann D, Krause C, Bach F, Windhagen H, Meyer-Lindenberg A (2010) Degradation behaviour and mechanical properties of magnesium implants in rabbit tibiae J Mater Sci 45: 624–632.

[41] Seitz J, Collier K, Wulf E, Bormann D, Bach F (2011) Comparison of the corrosion behavior of coated and uncoated magnesium alloys in an in vitro corrosion environment Adv. Eng. Mater. 13: B313-B323.

[42] Jha AM, Singh AC (1994) Clastogenicity of lanthanides--induction of micronuclei in root tips of Vicia faba Mutat. Res. 322: 169–172.
[43] Hirano S, Suzuki KT (1996) Exposure, metabolism, and toxicity of rare earths and related compounds Environ. Health Perspect. 104: 85–95.
[44] Wells WH, Wells V (2001) The Lanthanides, Rare Earth Metals. In: Bingham E, editor. Metals and metal compounds, compounds of inorganic nitrogen, carbon, oxygen, and halogens. New York: Wiley. pp. 423–458.
[45] Feyerabend F, Fischer J, Holtz J, Witte F, Willumeit R, Drücker H, Vogt C, Hort N (2010) Evaluation of short-term effects of rare earth and other elements used in magnesium alloys on primary cells and cell lines Acta Biomater 6: 1834–1842.
[46] Alexander Feldhaus (2006) Wirkung von Seltenen Erden auf den osteoporotisch veränderten Knochen im Tiermodell der ovariektomierten Ratte. Available: http://edoc.ub.uni-muenchen.de/6361/1/Feldhaus_Alexander.pdf
[47] Witte F, Kaese V, Haferkamp H, Switzer E, Meyer-Lindenberg A, Wirth CJ, Windhagen H (2005) In vivo corrosion of four magnesium alloys and the associated bone response Biomaterials 26: 3557–3563.
[48] Erne P, Schier M, Resink TJ (2006) The road to bioabsorbable stents: reaching clinical reality? Cardiovascular and interventional radiology 29: 11–16.
[49] Tschegg E, Lindtner RA, Doblhoff-Dier V, Stanzl-Tschegg SE, Holzlechner G, Catellani C, Imwinkelried T, Weinberg A (2010) Characterization methods of bone-implant-interfaces of bioresorbable and titanium implants by fracture mechanical means J Mech Behav Biomed Mater 4: 766-75
[50] Castellani C, Lindtner RA, Hausbrandt P, Tschegg E, Stanzl-Tschegg SE, Zanoni G, Beck S, Weinberg A-M (2011) Bone-implant interface strength and osseointegration: Biodegradable magnesium alloy versus standard titanium control Acta Biomater 7: 432–440.
[51] Lazar Leonovich Rokhlin (2003) Magnesium Alloys Containing Rare-earth Metals. London: Taylor & Francis Ltd.
[52] Abdel MP, Morrey ME, Grill DE, Kolbert CP, An K-N, Steinmann SP, Sanchez-Sotelo J, Morrey BF (2012) Effects of joint contracture on the contralateral unoperated limb in a rabbit knee contracture model: A biomechanical and genetic study J Orthop Res doi:10.1002/jor.22106
[53] Reifenrath J, Gottschalk D, Angrisani N, Besdo S, Meyer-Lindenberg A (2012) Axial forces and bending moments in the loaded rabbit tibia in vivo Acta veterinaria Scandinavica 54: 21.
[54] Zhang W, Wang W, Chen Q-Y, Lin Z-Q, Cheng S-W, Kou D-Q, Ying X-Z, Shen Y, Cheng X-J, Nie P-F, Li X-C, Rompis F, Huang H, Zhang H, Mu Z-L, Peng L (2012) Effect of calcium citrate on bone integration in a rabbit femur defect model Asian Pac J Trop Med 5: 310–314.
[55] Huehnerschulte TA, Angrisani N, Rittershaus D, Bormann D, Windhagen H, Meyer-Lindenberg A (2011) In vivo corrosion of two novel magnesium alloys ZEK100 and AX30 and their mechanical suitability as biodegradable implants Materials 4: 1144–1167.

[56] Ullmann B, Reifenrath J, Dziuba D, Seitz J, Bormann D, Meyer-Lindenberg A (2011) In vivo degradation behaviour of the magnesium alloy LANd442 in rabbit tibiae Materials 4: 2197–2218.

[57] Hampp C, Ullmann B, Reifenrath J, Angrisani N, Dziuba D, Bormann D, Seitz J-M, Meyer-Lindenberg A (2012) Research on the biocompatibility of the new magnesium alloy LANd442 - an in vivo study in the rabbit tibia over 26 weeks Adv. Eng. Mater. 14: B28–B37.

[58] Höh N von der, Bormann D, Lucas A, Denkena B, Hackenbroich C, Meyer-Lindenberg A (2009) Influence of different surface machining treatments of magnesium-based resorbable implants on the degradation behavior in rabbits Adv Eng Mater 11: B47-B54.

[59] Happel CM, Klose C, Witton G, Angrisani GL, Wienecke S, Groos S, Bach F-W, Bormann D, Männer J, Yelbuz TM (2010) Non-destructive, high-resolution 3-dimensional visualization of a cardiac defect in the chick embryo resembling complex heart defect in humans using micro-computed tomography: double outlet right ventricle with left juxtaposition of atrial appendages Circulation 122: e561-4.

[60] Lalk M, Reifenrath J, Rittershaus D, Bormann D, Meyer-Lindenberg A (2010) Biocompatibility and degradation behaviour of degradable magnesium sponges coated with bioglass - method establishment within the framework of a pilot study Mat Sci Eng Tech 41: 1025–1034.

[61] Kraus T, Fischerauer SF, Hänzi AC, Uggowitzer PJ, Löffler JF, Weinberg AM (2012) Magnesium alloys for temporary implants in osteosynthesis: In vivo studies of their degradation and interaction with bone Acta Biomater 8: 1230–1238.

[62] Vecchio KS, Zhang X, Massie JB, Wang M, Kim CW (2007) Conversion of sea urchin spines to Mg-substituted tricalcium phosphate for bone implants Acta Biomater 3: 785–793.

[63] Song G, Atrens A (2003) Understanding Magnesium Corrosion - A Framework for Improved Alloy Performance Adv Eng Mater 5: 837–858.

[64] Song G (2005) Recent progress in corrosion and protection of magnesium alloys Adv Eng Mater 7: 563–586.

[65] Huehnerschulte TA, Reifenrath J, Rechenberg B von, Dziuba D, Seitz J-M, Bormann D, Windhagen H, Meyer-Lindenberg A (2012) In vivo assessment of the host reactions to the biodegradation of the two novel magnesium alloys ZEK100 and AX30 in an animal model Biomedical engineering online 11: 14–42.

[66] Niinomi M (2008) Metallic biomaterials J Artif Organs 11: 105–110.

[67] Romani AM, Scarpa A (2000) Regulation of cellular magnesium Front Biosci 5: d720-734.

[68] Ullmann B, Reifenrath J, Seitz J, Bormann D, Meyer-Lindenberg A (n.d.) Influence of the grain size on the in vivo degradation behaviour of the magnesium alloy LAE 442 Acta Biomater submitted

Section 4

Mechanical Properties – Magnesium Alloys

Chapter 5

Thermal Stability and Mechanical Properties of Extruded Mg-Zn-Y Alloys with a Long-Period Stacking Order Phase and Plastic Deformation

Masafumi Noda, Yoshihito Kawamura, Tsuyoshi Mayama and Kunio Funami

Additional information is available at the end of the chapter

http://dx.doi.org/10.5772/48202

1. Introduction

Lightweight Mg alloys with excellent shock-absorption properties are being actively adopted for use in electronic information devices and automotive parts [1]. For such structural applications, Mg alloys need to have adequate ductility, thermal stability, and strength. However, Mg alloys often exhibit low ductility and low tensile yield strength at room temperature and above as a result of a scarcity of slip systems in their hexagonal close-packed structures [2]. Effective ways of improving the ductility and tensile yield strength of Mg alloys include grain refinement [3] and control of the texture [4]; these techniques promote prismatic slips and facilitate the creation of large plastic deformations. Recently, alloys of Mg with transition metals (TMs), such as Co, Ni, Cu, or Zn, and rare-earth (RE) metals, such as Y, Gd, Tb, Dy, Ho, or Er, have been found to show superior mechanical properties to those of other Mg alloys [5,6]. A characteristic of these Mg–TM–RE alloy systems is the formation of a long-period stacking order (LPSO) phase in as-cast materials and/or after heat treatment. In the present study, we examined the annealing properties, tensile properties, thermal stability, and rolling workability of high-strength extruded $Mg_{96}Zn_2Y_2$ alloys.

$Mg_{96}Zn_2Y_2$ alloy contains an LPSO phase as a secondary phase in the dominant α-Mg phase [5,6]. In general, Mg alloys with LPSO phases are known to have greatly enhanced mechanical properties, whereas their ductility can be maintained only by extrusion and/or plastic deformation treatments of the cast metal. It has been suggested that kink deformations in the LPSO phase and microstructural refinement in the α-Mg phase occur during extrusion deformation [7]. The tensile yield strength, microstructure, fatigue properties, and thermal stability of extruded $Mg_{96}Zn_2Y_2$ and LPSO-type Mg alloy have been reported [8–10]. Enhancements in the strength of $Mg_{96}Zn_2Y_2$ alloy by means of extrusion have been investigated and reported by Kawamura *et al.* [6–7]. However, improvements in ductility, which are vital for industrial application of Mg alloys, have not been investigated in relation to the annealing properties and thermal stability of Mg–Zn–Y alloys. It was therefore necessary to investigate the relationship between changes in the microstructure of the α-Mg and LPSO phases and the reduction in strength and the improvement in ductility produced by annealing of the alloy. Although LPSO-type Mg alloys have a high strength, they also need to display good ductility and high thermal stability before they can find practical industrial applications. In the present study, we examined the effects of various annealing treatments on the microstructure, tensile yield strength, ductility, and thermal stability of samples of $Mg_{96}Zn_2Y_2$ alloy produced by extrusion at 623 K. Annealing was accomplished by holding the extruded sample for 1 to 100 h in an electric furnace at various temperatures between 473 and 773 K.

Additionally, many of the plastic deformation processes of LPSO-type Mg alloy and other high-strength materials are affected by extrusion processing and there have been few attempts to examine the development of high strength by rolling and other working processes. It is generally known that plastic deformation of Mg alloys containing added RE elements requires many working cycles and a high deformation temperature in comparison with commercial Mg alloys [6,11]. The use of wrought metals as industrial materials is important and, to reduce costs, it is desirable to use plastic-deformation as well as extrusion in processing the materials. In the case of Mg alloys, the establishment of a plastic deformation process that combines a heat-treatment process, a plastic-working process, and controlled microstructures, leading to a reduction in processing costs, is urgently required. We produced a time–temperature–transformation (*TTT*) diagram for $Mg_{96}Zn_2Y_2$ alloys by investigating the microstructure, strength, elongation, and thermal stability of as-extruded and annealed samples at room temperature and at high temperatures. As a result, we succeeded in producing high-strength rolled sheet of $Mg_{96}Zn_2Y_2$ by a few passes of a rolling processing to control the amounts of LPSO phase and α-Mg phase. The fine-grained α-Mg phase did contribute to plastic deformation, but the grain size in the rolled sheet was 5 μm, suggesting that the LPSO phase was responsible for flexible workability as a result of the presence of continuous bending and kink bands at the boundary between the two phases.

2. Experimental procedures

The alloys used in this study were extruded specimens of $Mg_{96}Zn_2Y_2$ (atomic %) alloys. Ingots were prepared by high-frequency induction melting in an Ar atmosphere. A 60-mm-long cast sample with a diameter 29 mm was prepared. This $Mg_{96}Zn_2Y_2$ ingot was extruded at 623 K at an extrusion ratio of 10 and an extrusion speed of 2.5 $mm \cdot s^{-1}$. The $Mg_{96}Zn_2Y_2$ alloy consisted of an α-Mg phase, an LPSO phase, and inclusions of $Mg_3Zn_3Y_2$ compounds, with the α-Mg and LPSO phases as the major phases. Samples for tensile testing with gauge sections 2.5 mm in width and 15 mm in length were machined in the direction parallel to the direction of extrusion from samples of material that had been extruded and annealed.

Tensile tests were carried out at an initial strain rate of 5.0×10^{-4} s^{-1} at temperatures between room temperature and 573 K. Hardness tests were carried out for 20 s at a load of 1.96 kN on a plane normal to the direction of extrusion by using a Vickers hardness testing machine. Annealing was carried out at temperatures between 473 and 773 K in an electric furnace for various times between 1 and 100 h, with a maximum holding time of 1000 h, and the specimens were subsequently cooled in water. The microstructures of the extruded and annealed samples were observed by optical microscopy (OM), scanning electron microscopy (SEM), transmission electron microscopy (TEM), and electron backscattering diffraction (EBSD). Samples for microscopy were prepared by using a section polisher. In this study, cross-sections of the microstructure were observed from the direction of the extrusion axis. The SEM micrographs of cast and extruded materials are shown in Figure 1. These micrographs were recorded on cross-sectional planes perpendicular to the direction of extrusion. The brighter areas in Figure 1 correspond to the LPSO phase. The LPSO phase appears to be more finely distributed in the extruded alloys than in the as-cast alloys. The remaining 30% of the α-Mg phase region consisted of a nonrecrystallized material. The volume fraction of the LPSO phase was about 25%.

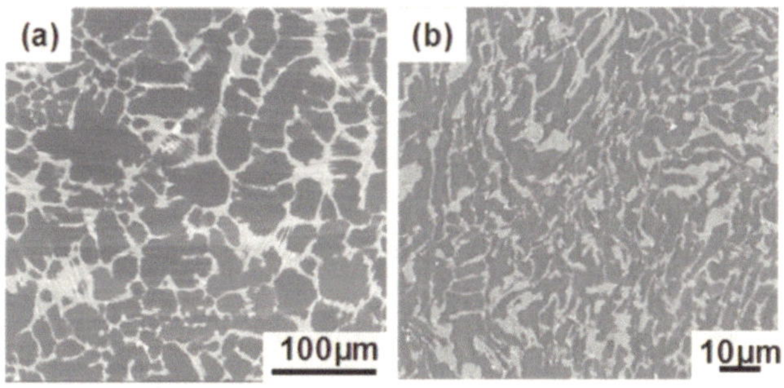

Figure 1. SEM micrographs of cast and extruded samples of $Mg_{96}Zn_2Y_2$ alloy.

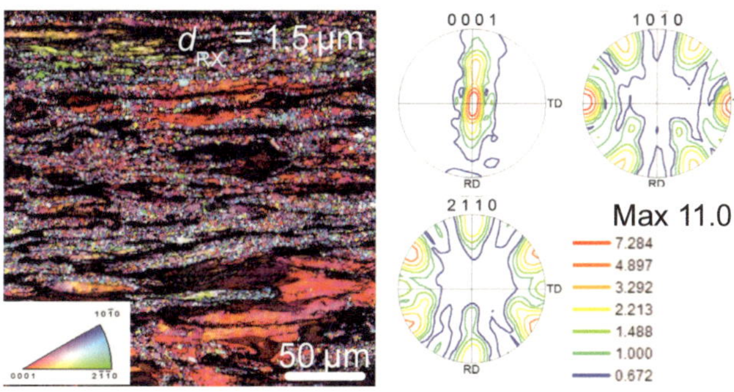

Figure 2. IPF and PF maps of as-extruded samples of $Mg_{96}Zn_2Y_2$. These maps were determined by EBSD, and the intensity of texture is indicated in the PF map.

After we had investigated the microstructure and tensile properties of the as-extruded and annealed materials at room temperature and at high temperatures, we investigated the hot-rolling workability of the material to elucidate the effect of the α-Mg phase and the LPSO phase on the plastic deformability of the $Mg_{96}Zn_2Y_2$ alloy.

In the rolling process, we used extruded $Mg_{96}Zn_2Y_2$ alloy in the form of the as-received material. The α-Mg phase showed static recrystallization grain growth. Kink bands in the LPSO phase were restored by the annealing treatment. Samples 5-mm thick, 20-mm wide, and 50-mm long were machined from the annealed and extruded materials and subjected to rolling in the direction perpendicular to the direction of extrusion. The rolling process was performed at a roll temperature of 473 K and a roll speed of 0.17 m s^{-1}. The sample was subsequently heated at 643 K for 5 min in an electric furnace and, finally, the rolled samples were quenched in water. Tensile samples with a gauge section 2.5 mm in width, 1 mm in thickness, and 12 mm in length were machined in the direction parallel to the rolling direction from the rolled sheet and from the various annealed sheets. In the as-rolled sheet, the cross section of the microstructure was observed from the rolling direction. The methods used to investigate the microstructures and tensile properties of as-rolled $Mg_{96}Zn_2Y_2$ alloy were the same as those used for the extruded and annealed $Mg_{96}Zn_2Y_2$ alloys.

3. Results and discussion

3.1. Tensile properties at room temperature and at high temperatures of extruded and annealed samples of $Mg_{96}Zn_2Y_2$ alloy

The results of tensile tests performed at room temperature and at high temperatures on extruded $Mg_{96}Zn_2Y_2$ samples annealed at various temperatures with holding times of 1 h and 100 h are shown in Figure 3. The yield stress and the tensile strength of the as-extruded sample were 391 and 432 MPa, respectively, and its elongation was 5.4% at room temperature. When the sample was annealed at 623 K for 1 h, the yield stress was 380 MPa and the tensile strength was 410 MPa; annealing at higher temperatures, resulted in a gradual decrease in strength. Even when the annealing time was extended to 100 h, the same tendency of showing a gradual decrease in these properties was observed. In other words, the α-Mg phase undergoes a static recrystallization at 623 K. On the other hand, it is known that the LPSO phase is responsible for local recovery and/or static recrystallization at an annealing temperature of 673 K [10]. At an annealing temperature at 673 K, the strength deteriorated as a result of recovery of kink bands in the LPSO phase and grain growth of the α-Mg phase, but the elongation improved. It has been reported that annealing of conventional Mg alloys at 523 K for 1 h disrupts the crystalline orientation and causes a reduction in strength [12]. Therefore, in comparison with conventional Mg alloy, the ductility of $Mg_{96}Zn_2Y_2$ alloy can be improved while maintaining a high strength, despite the fact that $Mg_{96}Zn_2Y_2$ has a low ductility at room temperature. The hardness of the samples, the mean grain size of the α-Mg phase, and the size of the $Mg_3Zn_3Y_2$ compounds are shown in Figure 4 for each annealing temperature.

As in the case of tensile tests, annealing at 623 K caused almost no reduction in hardness, which fell from HV 101 before annealing to HV 100 after annealing. The mean grain size was 1.5 μm in the as-extruded sample, which still contained nonrecrystallized regions. On the other hands, in the extruded sample annealed at 623 K, the mean grain size was 2.5 μm owing to the recovery of recrystallized and nonrecrystallized regions or to static recrystallization, and the average size of inclusions of $Mg_3Zn_3Y_2$ compound was 0.4 μm. From Figures 3 and 4, the static recrystallization temperature of the α-Mg phase of this alloy was estimated to be 623 K or higher. When the alloy was annealed at 773 K, its yield stress and tensile strength fell to 210 and 320 MPa, respectively, and the hardness was HV 75, whereas the elongation was improved to 23%.

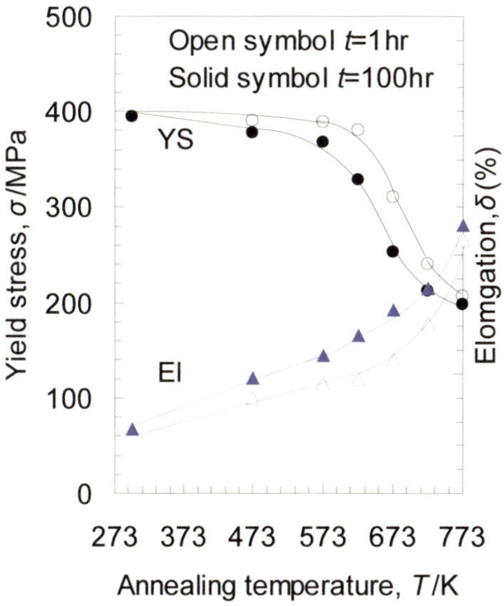

Figure 3. Mechanical properties of extruded and annealed samples of extruded samples of $Mg_{96}Zn_2Y_2$ alloy. Test temperatures: room temperature and 523 K; initial strain rate: 5.0×10^{-4} s^{-1}. T: testing temperature; T_a: annealing temperature; t_a: annealing time.

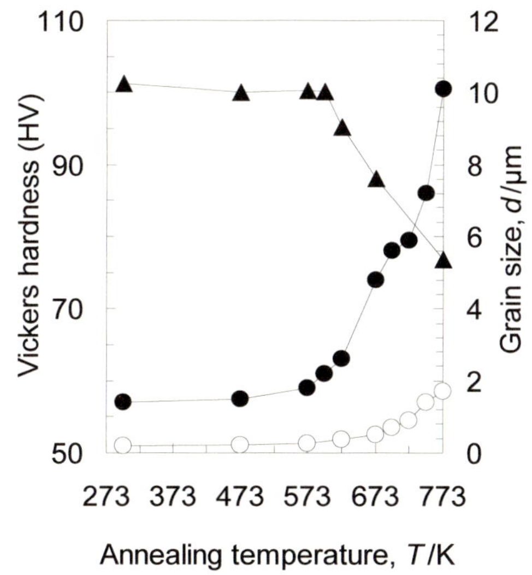

Figure 4. Changes in the hardness, recrystallized grain size of the α-Mg phase, and size of $Mg_3Zn_3Y_2$ phase caused by annealing.

The mean grain size of the α-Mg phase was about 10 μm, and the size of the inclusions of $Mg_3Zn_3Y_2$ compound was 1.7 μm. Fine-grained AZ31 Mg alloy annealed at 773 K for 1 h has a yield stress of 50 MPa, a mean grain size as large as 20 μm, and an improved elongation of 15% [13]. The alloy investigated in the present study showed even better mechanical properties. This was probably because the LPSO phases were dispersed within the α-Mg phase (even at 773 K), the mean grain size of the α-Mg phase was only about 10 μm, and the size of inclusions of $Mg_3Zn_3Y_2$ compound remained small. This tendency persisted when the annealing time was extended from 1 to 100 h, as demonstrated by the microstructural and tensile properties of the samples annealed for 100 h (see Section 3.2). We also carried out high-temperature tensile tests to investigate the temperature dependence of the properties of as-extruded and annealed $Mg_{96}Zn_2Y_2$ alloys; the results of these tests are shown in Figure 5. The tests were performed at 423 to 523 K at an initial strain rate of 5×10^{-4} s^{-1}. The yield stress and tensile strength of the as-extruded sample were 281 and 307 MPa, respectively, and the elongation was 24% at 523 K. The yield stress and elongation of extruded $Mg_{96}Zn_2Y_2$ alloy depended on the testing temperature, and the yield stress decreased with increasing testing temperature.

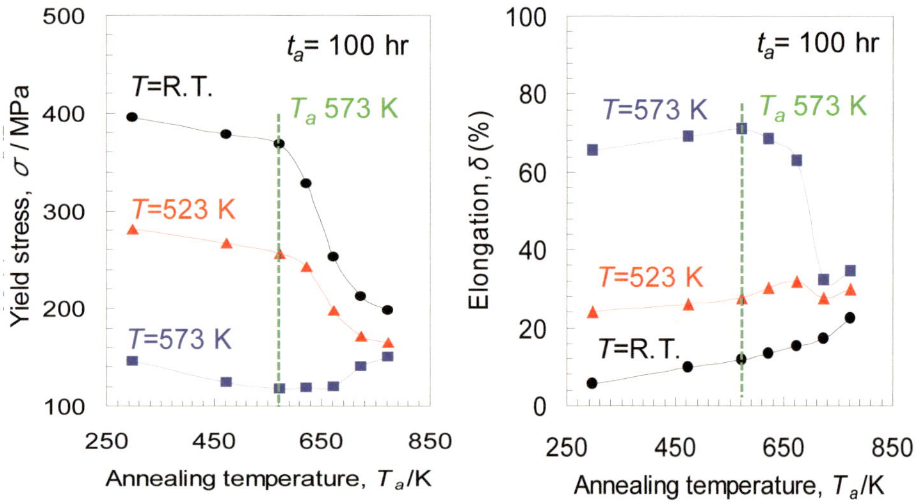

Figure 5. Mechanical properties of samples of $Mg_{96}Zn_2Y_2$ alloy extruded and annealed for 100 h at several temperatures. Tensile test were performed at room temperature to 573 K at an initial strain rate of 5.0×10^{-4} s^{-1}.

However, the dependence on the testing temperature was small in comparison with that of conventional Mg alloys [14]. In addition, the yield stress depended on the annealing temperature, but materials annealed at 573 K showed the same ratio as the yield stress of the as-extruded materials. There was a relative decrease in the yield stresses of materials annealed at a higher temperature of 673 K, regardless of the testing temperature. In this case, static recrystallization of the α-Mg phase and locally recovery and/or recrystallization of the LPSO phase of $Mg_{96}Zn_2Y_2$ alloy occur at 623 and 673 K, respectively. It is therefore important to maintain the kink band formed in the LPSO phase by extrusion to retain a high strength at high temperatures. When the tests were performed at 523 K, the yield stresses were 300 MPa for the as-extruded samples and 260 MPa for samples annealed at 623 K for 100 h. However, when the testing temperature was 573 K, the yield stress decreased to 120 MPa, regardless of the annealing temperature and time; however, the elongation showed a marked improvement to 70%. In other words the $Mg_{96}Zn_2Y_2$ alloy has excellent thermal stability and good mechanical properties for an annealing temperature of 573 K and a testing temperature of 523 K.

Figure 6 shows the effects of the annealing temperature and the annealing time on the mean recrystallized grain size of the α-Mg phase. The mean grain size of α-Mg phase increased with increasing annealing temperature, but it remained only 20 μm, even after annealing for 100 h at 773 K, because the presence of the LPSO phase restrained the growth of grains of the α-Mg phase. Furthermore, the α-Mg phase showed grain growth for annealing times for 10 h, but increase in grain size was gradual even if the annealing time was extended to 100 h. The grain size of the α-Mg phase remained at 10 μm for an annealing temperature of 673 K, suggesting that the LPSO phase did recover or recrystallize locally, or that the microstructure was stable.

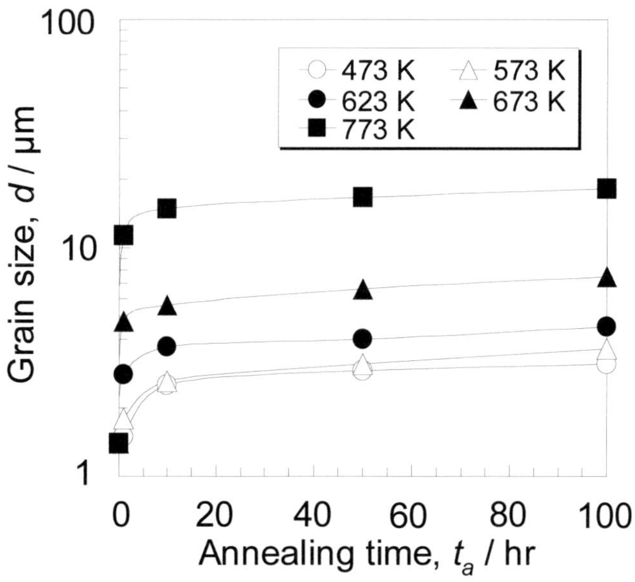

Figure 6. Relationship between the annealing conditions and the grain size of the α-Mg phase in samples of $Mg_{96}Zn_2Y_2$ alloy.

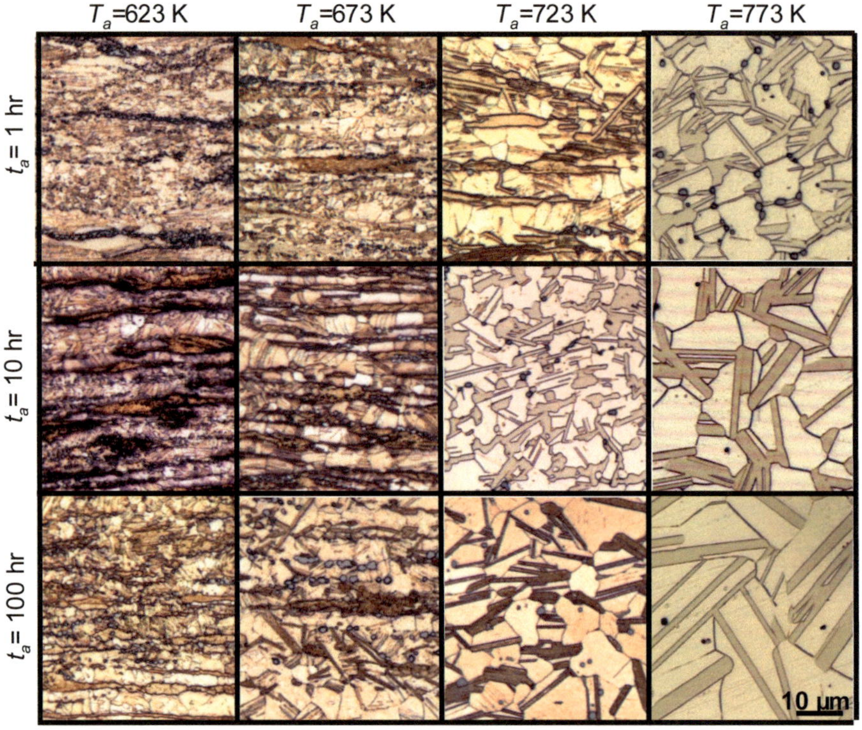

Figure 7. Optical micrographs of extruded samples of $Mg_{96}Zn_2Y_2$ alloy annealed at various temperatures for various holding times.

3.2. Microstructures of extruded and annealed samples of $Mg_{96}Zn_2Y_2$ alloy

Figure 7 shows optical micrographs of samples annealed at various temperatures and for various holding times. The average grain size of the α-Mg phase is shown in Figure 6. About 70% of the extruded $Mg_{96}Zn_2Y_2$ samples consisted of a recrystallized region and the remainder consisted of nonrecrystallized material. We suggest that discontinuous dynamic recrystallization occurs during extrusion. The grain size of the α-Mg phase increased on increasing the annealing temperature, but the grain size and the proportion of the recrystallized region did not change markedly at a temperature of 573 K. SEM micrographs of samples annealed at 623 K for 1 and 100 h are shown in Figure 8. The SEM micrographs show that, whereas the LPSO phase formed a block-type structure after annealing at 623 K for 1 h, a plate-type LPSO phase additionally appeared when the annealing time was extended to 100 h; the form of the LPSO phase therefore changed from a block type to a plate type when the annealing temperature was increased. Although annealing for about 1 h resulted only in local recovery or recrystallization of kink bands formed in the LPSO phase [10,15], an extension of the annealing time permitted a change in form of the LPSO phase, even at 623 K. The formation of the plate-type LPSO phase from the block-type one is considered to result from redissolution and precipitation of the constituent elements.

SEM micrographs, TEM micrographs, IPF maps, and PF maps of samples annealed at 573 and 623 K for 100 h are shown in Figure 9. The SEM micrographs show that a plate-type LPSO phase appeared in samples annealed at 623 K for 100 h. The mean grain size of the α-Mg phase of materials annealed at 573 and 623 K were 3.4 and 4.8 μm, respectively, the intensities of texture were 6.1 and 3.0, respectively, and the proportions of recrystallized regions were 83 and 91%, respectively. The reason why the material annealed at 573 K show a high strength at room temperature and at high temperatures is that kink bands remain in the block-type LPSO phase and because the static recrystallization temperature of $Mg_{96}Zn_2Y_2$ alloy is higher than that shown by Mg alloys in general. As can be seen in Figure 9, the intensity of texture of samples annealed at 573 K for 100 h was similar to that of a rolled sheet of AZ31 Mg alloy subjected to working from 473 to 673 K until the rolling reduction reached 85.7%; the values were 7 and 5, respectively [16]. In other words, the presence of an LPSO phase containing kink bands is more important than are microstructural changes in the α-Mg phase. When the annealing temperature was increased to 673 K, the mean grain size of the α-Mg phase became 4.8 μm and kink bands showed a partial recovery, as observed by TEM studies on the microstructure (Figure 9). Figures 3–8 also show that significant changes occur in the mechanical properties and in the microstructure of $Mg_{96}Zn_2Y_2$ alloy during annealing at temperatures above 623 K. The recovery temperature for the LPSO phase is about 50 K higher than that of the α-Mg phase. It is important to note that formation of a kink band in the LPSO phase and microstructural refinement of the α-Mg phase during plastic deformation are responsible for the improvements in the mechanical properties of the alloy. In addition, we concluded that the alloy has superior strength and ductility because the LPSO phase constrains the growth grains in the α-Mg phase during annealing.

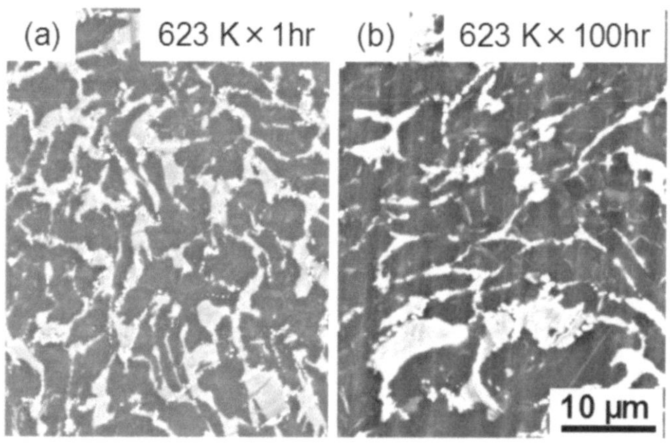

Figure 8. SEM micrographs of samples of $Mg_{96}Zn_2Y_2$ alloy annealed for (a) 623 K for 1 h and at (b) 623 K for 100 h.

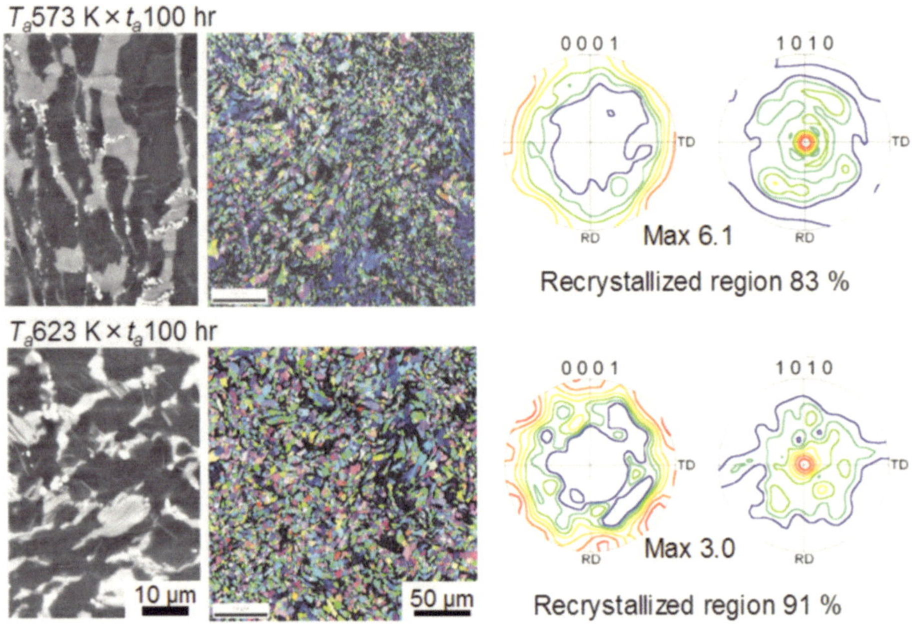

Figure 9. SEM micrographs, IPF maps, TEM micrographs, and PF maps of samples of $Mg_{96}Zn_2Y_2$ alloy annealed at 573 or 623 K for 100 h.

3.3. Thermal stability and the time–temperature–transformation diagram of extruded and annealed samples of $Mg_{96}Zn_2Y_2$

The extruded samples of $Mg_{96}Zn_2Y_2$ alloy did not shows any marked loss of strength or change in microstructure for annealing conditions of 573 K for 100 h and a testing temperature of 523 K. We therefore extended the annealing time to 1000 h to test the thermal stability of the alloy. Figure 10 shows the yield stress and elongation for a sample of extruded $Mg_{96}Zn_2Y_2$ alloy annealed at 573 K for various times up to 1000 h, as tested at room temperature and at 523 K. The yield stress and the tensile strength decreased gradually with increasing annealing time, whereas the elongation markedly improved. The tensile properties of samples annealed at 573 K for 1000 h did not depend on the annealing time or the testing temperature. Figure 11 shows SEM and TEM micrographs and an IPF map of a sample annealed at 573 K for 1000 h. Figure 10 shows that when the annealing temperature is maintained at 573 K for 1000 h, even though the yield stress is reduced from 391 to 352 MPa, the rates of decrease remain at 10 and 7%, respectively, and elongation is improved from 8 to 11%. Although the α-Mg phase grows from 1.5 μm to 4.8 μm, the LPSO phase forms a block-type structure, and a lamellar microstructure in which the LPSO phase is finely dispersed in the α-Mg phase forms; no substantial change was observed, even after annealing for 1000 h. In comparison with the TEM microstructure shown in Figure 11, where the sample was annealed for 1000 h, the LPSO phase is more bent and kink bands remain. The lamellar microstructure of the α-Mg and LPSO phases were the same as those observed before annealing.

The grains of the α-Mg phase coarsen as the annealing time is increased (see Figures 6 and 11). However, because the LPSO phase inhibits grain growth, the grain size is still fine compared with that of commercially available alloys. A crystallographic orientation analysis showed that 63% of the measured view for extruded materials and for materials annealed for 1 h consisted of high-angle grain boundaries and that the proportion of {10-10} crystalline orientation was 38%. However, because the grain size of the α-Mg phase increases with increasing annealing time, eventually 78% of the measured view consisted of high-angle grain boundaries and the proportion of {10-10} crystalline orientation fell to 11%. In other words, degradation of the mechanical properties of the extruded materials after annealing at 573 K is due to static recovery of the alloys. With regard to the reinforcing factor of this material, kink bands remained in the LPSO phase even after annealing at 573 K for 1000 h, and the LPSO phase formed a block-type structure. By the way, it is known that the LPSO phase is a harder material than the α-Mg phase. When an LPSO phase is present, the Mg alloy has a high strength and the grain growth of the α-Mg phase is constrained, but the microstructure remains unchanged. In the case of Mg alloys, it is important that they retain a high strength and a high ductility at high temperatures if they are to be used as industrial materials, and the LPSO-type Mg alloy is a material that can solve many such problems.

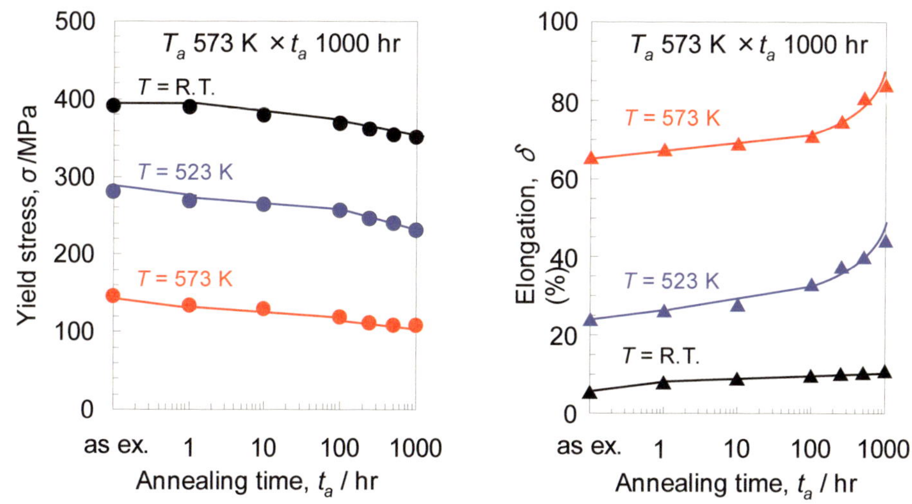

Figure 10. Relationship between the annealing time, mechanical properties, and testing temperatures for extruded samples of $Mg_{96}Zn_2Y_2$ alloy annealed at 573 K.

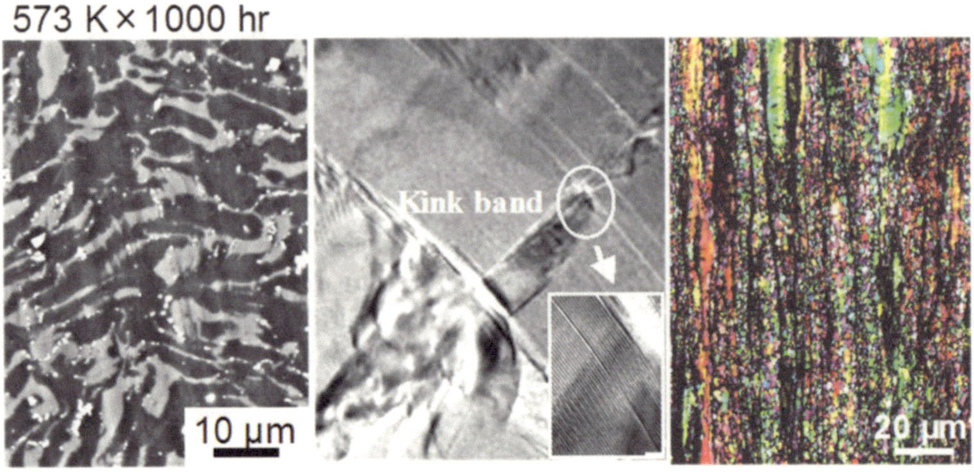

Figure 11. SEM and TEM micrographs and an IPF map of a sample of $Mg_{96}Zn_2Y_2$ alloy annealed at 573 K for 1000 h.

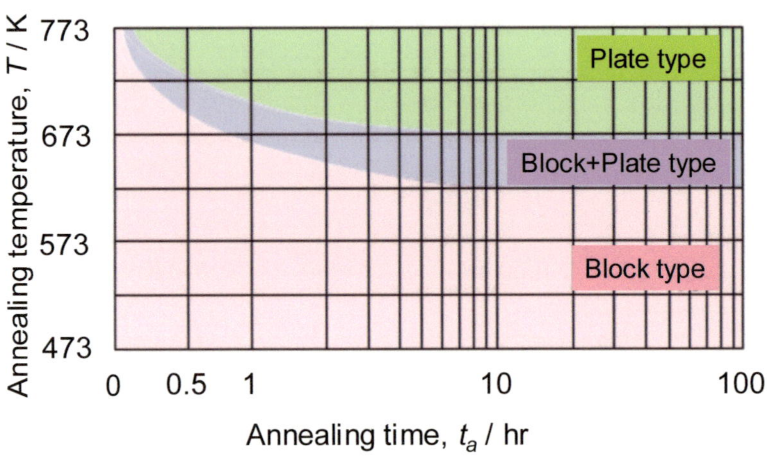

Figure 12. *TTT* diagram for extruded $Mg_{96}Zn_2Y_2$ alloy.

The time–temperature–transformation (*TTT*) diagram of $Mg_{96}Zn_2Y_2$ alloys (Figure 12), obtained by investigating the microstructure, strength, elongation, and thermal stability of as-extruded and annealed samples at room temperature and high temperatures, can be used to optimize working and forming processes. In Figure 12, the annealing temperature and annealing time graph is divided into three regions, depending on the shape of the LPSO phases: block type, block + plate type, or plate type. In the block-type region, annealing for a long time does not cause a marked deterioration in heat resistance or high strength. On the other hands, the plate-type region provides a high ductility, although the strength is reduced. We chose this classification in terms of the shape of LPSO phase because kink bands introduced into the LPSO phase play an important role in relation to the expression of high strength and high thermal stability of the $Mg_{96}Zn_2Y_2$ alloy.

3.4. Strength, microstructure, anisotropy, and rollability of samples of $Mg_{96}Zn_2Y_2$ alloy

Figure 13 shows SEM micrographs of an as-extruded sample of $Mg_{96}Zn_2Y_2$ alloy and of a sample annealed at 773 K for 1 h. The LPSO phase in the annealed materials appears to be more finely dispersed in a plate-type form than that in the extruded alloy. The volume fraction of the LPSO phase is similar to that of the as-extruded alloy, and a plate-type LPSO phase formed in several directions in the grains and grain boundaries. Normally, the LPSO phase, which is the secondary phase, acts as a fiber reinforcement and is anisotropic after plastic deformation. It has been reported [17] that in the as-extruded material, the yield stress in the direction of extrusion shows a difference of about 30% from that in the direction perpendicular to the extrusion direction. Here, the rolling process was performed in four passes at 643 K with a roll temperature of 473 K. The samples were subsequently heated at 643 K for 5 min in an electric furnace and finally quenched in water. Table 1 shows the anisotropy of the tensile properties of hot-rolled sheets of the alloy, where an angle of 0° is parallel to the rolling direction. The maximum value of the anisotropy of the yield stress of the rolled material was about 10%, which is less than one third of that of the extruded material.

Tensile direction	YS (MPa)	UTS (MPa)	Elongation (%)
0°	365	410	5.5
45°	350	402	12
90°	328	398	7

Table 1. Relationship between the mechanical properties and the tensile direction for a rolled $Mg_{96}Zn_2Y_2$ sheet. The tensile test direction 0° was parallel to the direction of rolling.

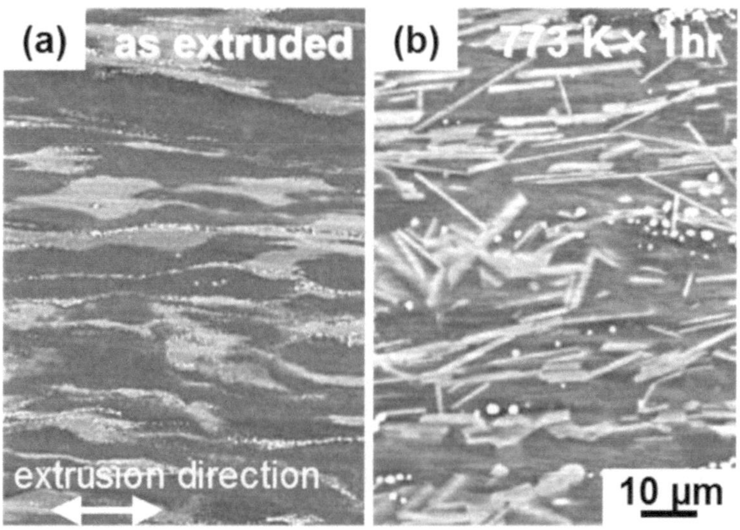

Figure 13. SEM micrographs of (a) as-extruded alloy $Mg_{96}Zn_2Y_2$ alloy and (b) a sample annealed at 773 K for 1 h.

The dispersal of the LPSO phase by heat treatment, as shown in Figure 13, is effective in reducing the anisotropy, and even low-deformation processing results in a high strength of over 360 MPa. The tensile strength was almost constant and was independent of the test direction. However, the elongation was maximal for a tensile direction of 45°. The LPSO phase is known to be a hard phase, but we were able to reduce the rolling processing temperature by about 30 K and to perform the processing by using a schedule consisting of only a few passes. A four-pass schedule was the minimum number of passes for rolling the LPSO-type Mg alloy, and the strain ratio introduced by the rolling processing was 1.8, which is less than that produced by the extrusion process. However, the 0.2% proof stress and elongation of $Mg_{96}Zn_2Y_2$ rolled sheet material were 360 MPa and 5.5%, respectively. In other words, the Mg alloy regained its high strength even when its LPSO phase was not subject to severe plastic deformation.

Optical micrographs and IPF maps are show in Figures 14 and 15, respectively. In Figure 14, the darker regions correspond to the LPSO phase. The LPSO phase has a large curvature in the rolling direction and shows considerable deformation. Furthermore, the formation of kink bands can be recognized in the LPSO phase. Normally, the LPSO phase does not show a large curvature in extruded materials, and we suggest that the large deformation of the LPSO phase observed in this case is responsible for the development of high strength by the rolling process. From Figure 15, we estimated that the grain size of the α-Mg phase in the rolled material was 5.0 μm. The intensity of texture of rolled sheet was 4.9, which is low in comparison with that of as-rolled AZ31 Mg alloy. Fine grains (mean size <5.0 μm) were formed around the LPSO phase, and it is possible that stress concentration acts on the α-Mg phase, because the LPSO phase is bent in the rolling process, causing dynamic recrystallization. The frequencies of low-angle and high-angle grain boundaries in the $Mg_{96}Zn_2Y_2$ rolled sheet were 19.3% and 80.7%, respectively. The rolling process is therefore capable of producing a high frequency of high-angle grain boundaries and a random crystal orientation in $Mg_{96}Zn_2Y_2$ alloy.

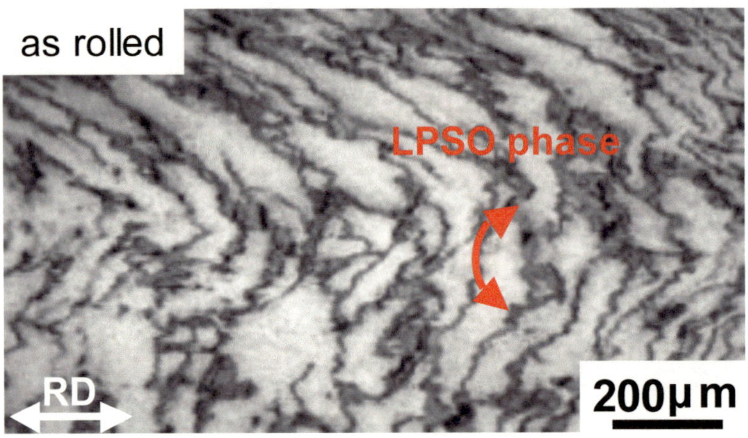

Figure 14. Optical micrograph of as-rolled $Mg_{96}Zn_2Y_2$ alloy.

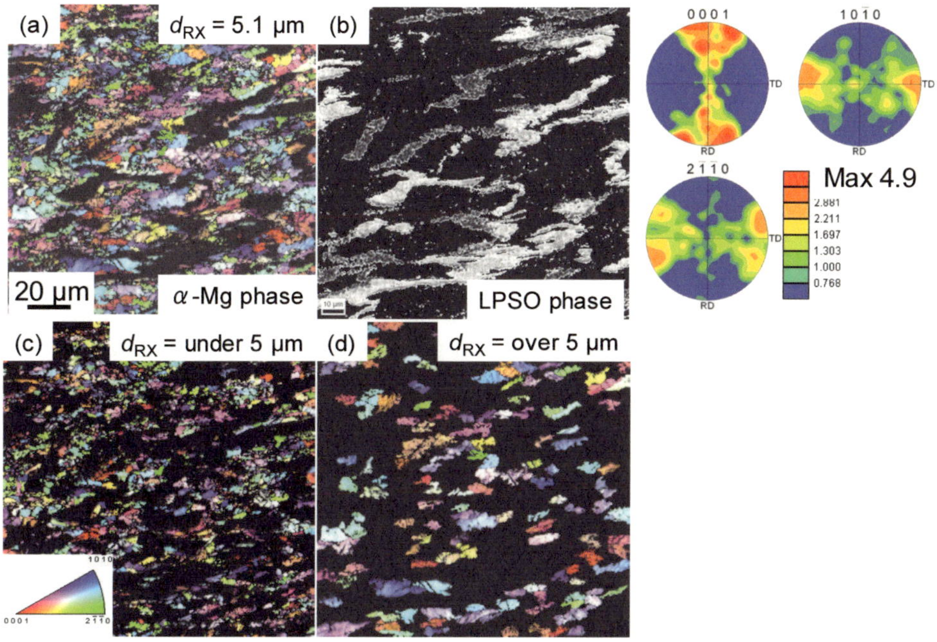

Figure 15. IPF and PF maps of as-rolled $Mg_{96}Zn_2Y_2$ alloy; the intensity of texture is indicated in the PF maps. Figures (c) and (d) were cropped from the IPF map; (c) shows the fine-grain region and (d) shows the coarse-grain region.

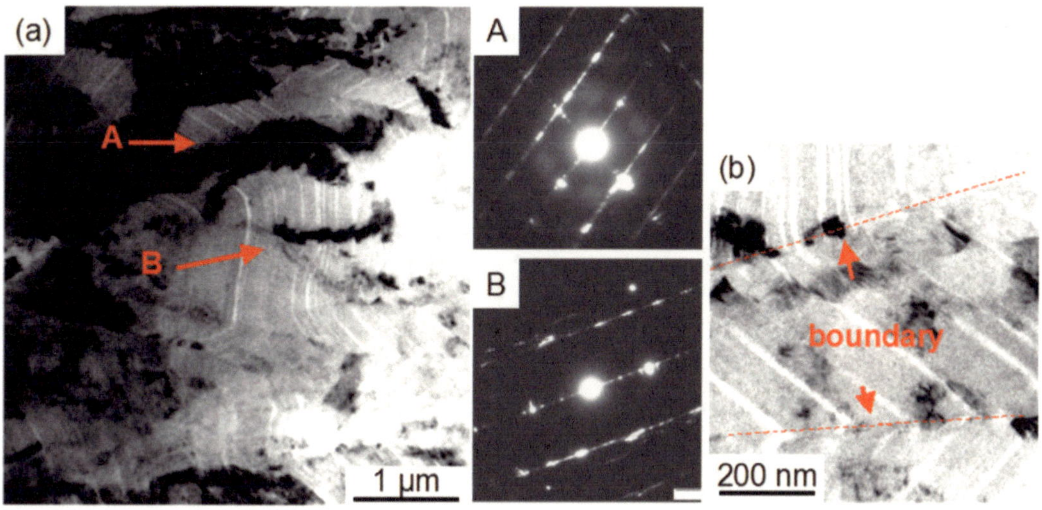

Figure 16. TEM micrographs of the as-rolled $Mg_{96}Zn_2Y_2$ alloy; (b) is high-magnification image of a kink boundaries.

TEM micrographs of the LPSO phases in an as-rolled sheet of $Mg_{96}Zn_2Y_2$ alloy are shown in Figure 16. In the rolled sheet, the LPSO phase was continuously deformed, but boundaries were observed in some regions, suggested that the LPSO phase recovered, at least partially, from the kink deformation introduced during rolling process. Although it is difficult to compare results of a previous study [18] directly with those of the present study, for FCC metals, cellular microstructures at the boundary between the deformation band and the deformation matrix act as recovery nuclei during subsequent reheating. We assume that the kinks acted as recovery nuclei in the $Mg_{96}Zn_2Y_2$ alloy and led to the formation of a static structure when edge dislocations accumulated as a result of reheating of the kink deformation that formed within the LPSO phase. In the $Mg_{96}Zn_2Y_2$ rolled sheet, the LPSO phase was continuously deformed, but boundaries were observed in some regions, suggesting that the LPSO phase recovered, at least partially, from the kink deformation introduced during the hot-rolling process. The present study clarified that continuous deformation of kink bands is necessary to permit the LPSO phase to show large deformations and that recovery of kink bands during reheating is important.

4. Conclusions

The mechanical properties and microstructures of extruded $Mg_{96}Zn_2Y_2$ alloys annealed at various temperatures with various holding times were investigated. Annealing the extruded $Mg_{96}Zn_2Y_2$ alloy at 623 K caused reductions in yield stress and a gradual decrease in the tensile strength; however, the ductility improved from 5.4 to 9.6%. The α-Mg phase and LPSO phase showed static recrystallization and recovery at 623 and 673 K, respectively; therefore the $Mg_{96}Zn_2Y_2$ alloy showed a superior thermal stability when it was annealed at 573 K. An effective way to improve the ductility of the alloy while maintaining high strength involves ensuring recovery of the secondary LPSO phase from kink bands, controlling the grain size of the α-Mg phase, and producing a fine dispersion of the LPSO phase in the α-Mg phases. In hot-rolled samples of $Mg_{96}Zn_2Y_2$ alloy, the anisotropy of the rolled sheet alloy was low compared with that of the extruded alloy when rolling was performed at 673 K after annealing treatment. The important factors in the development of $Mg_{96}Zn_2Y_2$ alloy with a high strength and ductility are a fine dispersion of the LPSO phase in the grain boundary and the formation of continuous kink bands and/or boundary in the LPSO phase.

Author details

Masafumi Noda* and Kunio Funami
Department of Mechanical Science and Engineering,
Chiba Institute of Technology, Tsudanuma, Narashino, Chiba, Japan

Yoshihito Kawamura and Tsuyoshi Mayama
Department of Material Science, Kumamoto University, Kurokami, Kumamoto, Japan

*Corresponding author

Acknowledgement

The authors are grateful for the financial supports from the Kumamoto Prefecture Collaboration of Regional Entities supported this work for the Advancement of Technological Excellence, Japan Science and Technology Agency. The examination of rolling process went by the furtherance from Grant-in-Aid for Young Scientists for (B) No. 23760099 from Japan Society for the Promotion of Science.

5. References

[1] Alan A. Luo (2003) Recent Magnesium Alloy Development for Automotive Powertrain Applications. Materials Science Forum 419-422: 57-66.
[2] M.H. Yoo (1981) Slip, Twinning, and Fracture in Hexagonal Close-Packed Metals. Metallurgical Transactions A 12A: 409-418.
[3] S.E. Ion, F.J. Humphreys, S.H. White (1982) Dynamic Recrystallisation and the Development of Microstructure During the High Temperature Deformation of Magnesium. Acta Metallurgica 30: 1909-1919
[4] H. Watanabe, K. Ishikawa (2009) Effect of Texture on High Temperature Deformation Behavior at High Strain Rates in a Mg-3Al-1Zn Alloy. Mater. Sci. Eng. A 523: 304-311.
[5] Y. Kawamura, K. Hayashi, A. Inoue, T. Masumoto (2001) Rapidly Solidified Powder Metallurgy Mg97Zn1Y2 Alloys with Excellent Tensile Yield Strength Above 600 MPa. Materials Transactions 42: 1172-1176.
[6] Y. Kawamura, M. Yamasaki (2007) Formation and Mechanical Properties of Mg97Zn1RE2 Alloys with Long Period Stacking Ordered Structure. Materials Transactions 48: 2986-2992.
[7] S. Yoshimoto, M. Yamasaki, Y. Kawamura (2006) Microstructure and Mechanical Properties of Extruded Mg-Zn-Y Alloys with 14H Long Period Ordered Structure. Materials Transactions 47: 959-965.
[8] T. Itoi, T. Inazawa, Y. Kuroda, M. Yamasaki, Y. Kawamura, M. Hirohashi (2010) Tensile property and cold formability of a Mg96Zn2Y2 alloy sheet with a long-period ordered phase. Materials Letters 64: 2277-2280.
[9] R. Matsumoto, M. Yamasaki, M. Otsu, Y. Kawamura (2009) Forgeability and Flow Stress of Mg-Zn-Y Alloys with Long Period Stacking Ordered Structure at Elevated Temperatures. Materials Transactions 50: 841-846.
[10] M. Noda, T. Mayama, Y. Kawamura (2009) Evolution of Mechanical Properties and Microstructure in Extruded Mg96Zn2Y2 Alloys by Annealing. Mater. Trans. 50: 2526-2531.
[11] N. Stanford (2010) Micro-Alloying Mg with Y, Ce, Gd and La for Texture Modification – A Comparative Study. Materials Science and Engineering A 527: 2669-2677.
[12] M.R. Barnett, Z Keshavarz, A.G. Beer, D. Atwell (2004) Influence of Grain Size on the Compressive Deformation of Wrought Mg-3Al-1Zn. Acta Materialia 52: 5093-5103.
[13] Su C.W., Lu L., Lai M.O. (2007) Mechanical Behavior and Texture of Annealed AZ31 Mg Alloy Deformed by ECAP. Materials Science and Technology 23: 290-296.

[14] W.J. Kim, S.W. Chung, C.S. Chung, D. Kum (2001) Superplasticity in Thin Magnesium Alloy Sheets and Deformation Mechanism Maps for Magnesium Alloys at Elevated Temperatures. Acta Materialia 49: 3337-3345.

[15] M.M. Myshlyaev, H.J. McQueen, A. Mwembela, E. Konopleva (2002) Twinning, Dynamic Recovery and Recrystallization in Hot Worked Mg-Al-Zn Alloy. Materials Science and Engineering A 337: 121-133.

[16] Y. Chino, M. Mabuchi, R. Kisihara, H. Hosokawa, Y. Yamada, C. Wen, K. Shimojima, H. Iwasaki (2002) Mechanical Properties and Press Formability at Room Temperature of AZ31 Mg Alloy Processed by Single Roller Drive Rolling. Material Transactions 43: 2554-2560.

[17] T. Mayama, M. Noda, Y. Kawamura, K. Hagihara (2009) Anisotropic Compressive Behavior of Extruded Magnesium Alloy Mg96Zn2Y2: Experimental Observation and Crystal Plasticity Analysis. Collected extended abstracts of Thermec 2009: 302.

[18] K. Higashida, J. Takamura, N. Narita (1986) The Formation of Deformation Bands in F.C.C. Crystals. Materials Science and Engineering 81: 239-258.

Section 5

Magnesium Alloys – Welding and Joining Processes

Chapter 6

Welding of Magnesium Alloys

Parviz Asadi, Kamel Kazemi-Choobi and Amin Elhami

Additional information is available at the end of the chapter

http://dx.doi.org/10.5772/47849

1. Introduction

Magnesium is the sixth most abundant element on the Earth's surface, with virtually inexhaustible supplies in the oceans. It is the third most plentiful element dissolved in seawater, with an approximate concentration of 0.14% (Busk, 1987). Over recent years the industrial output of magnesium alloys has been rising by almost 20% per annum. Magnesium and its alloys, as the lightest structural material, are about 40% lighter than aluminium and as much as about 78% lighter than steel. It is demonstrated that using magnesium alloys results in a 22–70% weight reduction, compared to using alternative materials (Kulekci, 2008). Magnesium alloys have excellent specific strength, excellent sound damping capabilities, good cast-ability, hot formability, excellent machinability, good electromagnetic interference shielding, and recyclability (Haferkamp et al., 2000), (Mordike and Ebert, 2001), (Pastor et al., 2000). Moreover, magnesium ignites with difficulty in air due to its high heat capacity. Some disadvantages of magnesium are presented based on the following, Low elastic modulus; High degree of shrinkage on solidification; High chemical reactivity. Additionally, these alloys have limited fatigue and creep resistance at elevated temperatures (Mordike and Ebert, 2001). Because of the hexagonal close-packed (HCP) crystal structure, magnesium alloys also have a limited ductility and cold workability at room temperature (Sanders et al., 1999). These alloys have about the same corrosion resistance in common environments as mild steel, but are less corrosion resistant than aluminium alloys (Busk, 1987). Thus, magnesium alloy usage has been limited due to its poor corrosion resistance and low ductility. In order to overcome these problems, new alloys, such as Mg-AZ91D, have been developed and have improved corrosion resistance (Munitz et al., 2001). The property profiles demanded by automobile and other large-scale potential users of magnesium have revealed the need for alloy development. Fig. 1 illustrates the different trends in alloy development depending on the main requirement. The major alloying elements are aluminium, zinc, thorium and rare earths. Aluminium is the most important alloying element in the ternary Mg–Al series, which comprises AZ (Mg–Al–Zn), AM (Mg–Al–Mn) and AS (Mg–Al–Si) alloys. There are two binary systems employing manganese and zirconium (Oates, 1996). It is also common to classify magnesium alloys into those for room and elevated temperature applications. Rare earth metals and thorium are the main alloying elements for high temperature alloys. Aluminium and zinc, added either singly or in combination, are the most common alloying elements for room temperature applications because at elevated temperatures the tensile and creep properties degrade rapidly (Marya et al. 2000).

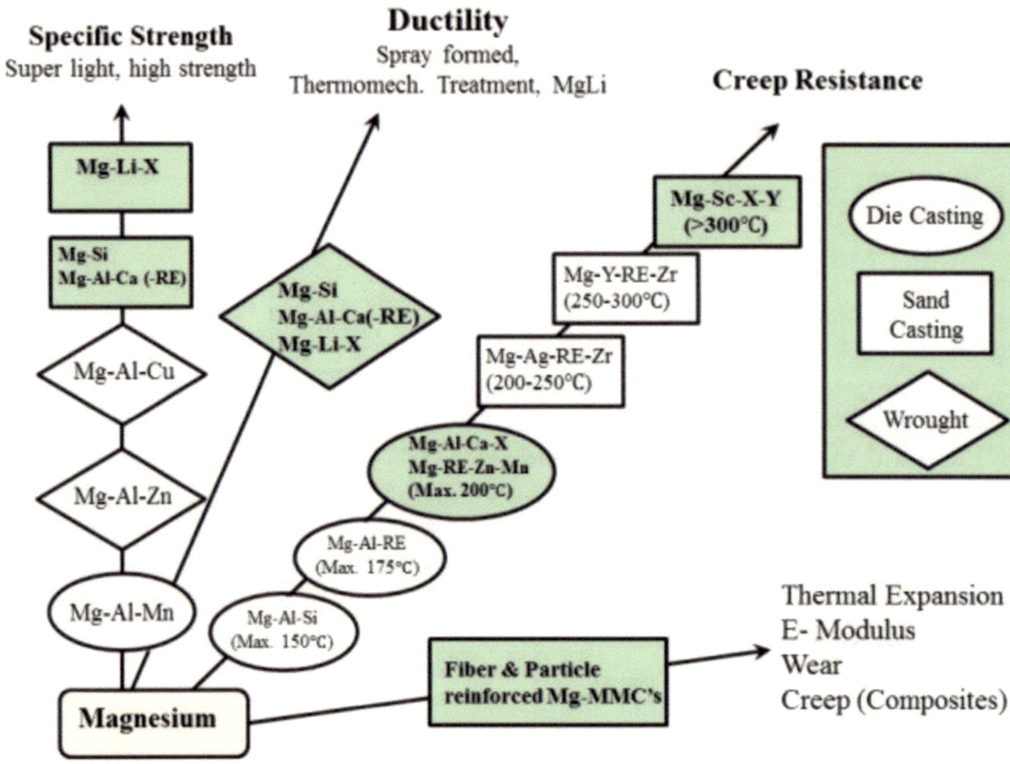

Figure 1. Directions of alloy development (Mordike and Ebert, 2001).

To date, no international code for designating magnesium alloys exists but the method used by the American Society for Testing Materials (ASTM) has been widely adopted. In this system, the first two letters indicate the principal alloying elements according to the following codes: A, aluminium; B, bismuth; C, copper; D, cadmium; E, rare earths; F, iron; G, magnesium; H, thorium; K, zirconium; L, lithium; M, manganese; N, nickel; P, lead; Q, silver; R, chromium; S, silicon; T, tin; W, yttrium; Y, antimony and Z, zinc. The two or one letter is followed by numbers which represent the nominal compositions of these principal alloying elements in weight percentage, rounded off to the nearest whole number. For example, AZ91 indicates the alloy Mg–9Al–1Zn with the actual composition ranges being 8.3–9.7 Al and 0.4–1.0 Zn. Suffix letters A, B, C, etc. refer to variations in composition within the specified range and X indicates that the alloy is experimental. Despite the good castability of some mentioned alloys (such as Mg-AZ91D alloys), it is not always possible or economically favourable to cast complex magnesium parts. Joining of magnesium parts, which may be crucial for these applications, is still restricted. It is, thus, very desirable that joining technologies be developed and made accessible for industrial applications (American Society of Metals [ASM], 1990).

Welding and joining processes are essential for the development of practically every manufactured product. However, these processes often appear to consume greater fractions of the product cost and to create more of the production difficulties than might be expected (ASM, 1993). There are a number of reasons that explain this situation. Because there are many fusion welding processes, one of the greatest difficulties for the manufacturing engineer is to define which process will produce satisfactory properties at the lowest cost. There are no simple answers. Any change in the part geometry, material, value of the end product, or size of the production run, as well as the availability of joining equipment, can influence the choice of joining method. For small lots of complex parts, fastening may be preferable to welding, whereas for long production runs, welds can be stronger and less expensive.

To date magnesium alloys have not usually been welded except for some repaired structures because of the occurrence of defects such as oxide films, cracks, and cavities (Haferkamp et al., 2000). However, the broader application of magnesium alloys requires reliable welding processes. Magnesium alloy components may be joined using mechanical clasps as well as a variation of welding methods including tungsten arc inert gas (TIG), plasma arc welding, electron beam welding (EBW), laser beam welding (LBW), friction stir welding (FSW), explosion, electromagnetic welding, ultrasonic welding, and resistance spot welding (RSW).

2. Laser beam welding

Laser beam welding (LBW) uses a moving high-density (105 to 107 W/cm2) coherent optical energy source called a laser as the source of heat. "Laser" is an acronym for "light amplification by stimulated emission of radiation." The coherent nature of the laser beam allows it to be focused to a small spot, leading to high energy densities (ASM, 1993). Fig. 2 illustrates the schematic of the laser beam welding.

Advantages and Limitations of LBW are listed below (ASM, 1993):

- Light is inertialess (hence, high processing speeds with very rapid stopping and starting become possible).
- Focused laser light provides high energy density.
- Laser welding can be used at room atmosphere.
- Difficult to weld materials (such as titanium, quartz and etc) can be joined.
- Workpieces do not need to be rigidly held.
- No electrode or filler materials are required.
- Narrow welds can be made.
- Precise welds (relative to position, diameter, and penetration) can be obtained.
- Welds with little or no contamination can be produced.

- The heat affected zone (HAZ) adjacent to the weld is very narrow.
- Intricate shapes can be cut or welded at high speed using automatically controlled light deflection techniques.
- The laser beam can also be time shared.

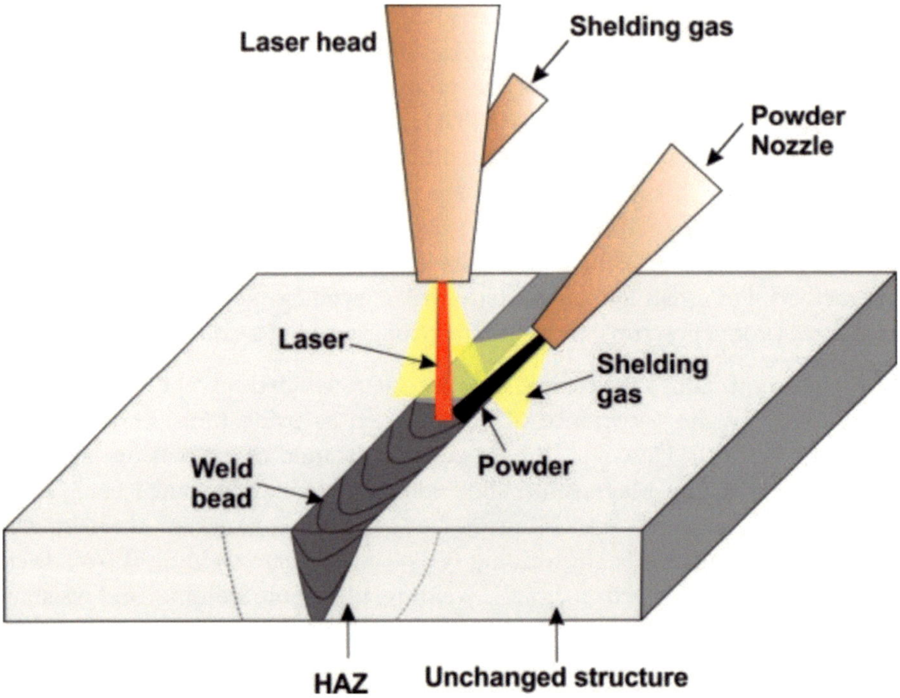

Figure 2. The schematic view of LBW process.

In the last two decades, the manufacturing industry has been actively engaged in qualifying laser welding, because high strength joints with low levels of residual stress and high visual quality can be achieved with this joining process. The effectiveness of laser welding depends greatly on the physical properties of the material to be welded. Magnesium alloys possess certain inherent characteristics such as low absorptivity of laser beams, strong tendency to oxidize, high thermal conductivity, high coefficient of thermal expansion, low melting and boiling temperatures, wide solidification temperature range, high solidification shrinkage, form low melting point constituents, low viscosity, low surface tensions, high solubility for hydrogen in the liquid state, and absence of colour change at the melting point temperature. Therefore, some processing problems and weld defects can be observed in laser welding of magnesium alloys such as unstable weld pool, substantial spatter (Haferkamp et al., 2001), strong tendency to drop through for large weld pools, sag of the weld pool (especially for thick workpiece), undercut (Lehner and Reinhart, 1999), porous oxide inclusions, loss of alloying elements, excessive pore formation (particularly for die castings) (Zhao and DebRoy, 2001), liquation and solidification cracking (Marya and Edwards, 2000). Despite the mentioned problems, among the welding technologies, laser welding has been considered to be an attractive and preferred fusion process due to high welding speed, very narrow joints with smaller heat affected zone (HAZ) because of using shielding gases, low distortion and excellent environment adaptability (Wang et al., 2001).

2.1. CO_2 and Nd:YAG lasers

Lasers have been promoted as potentially useful welding tools for a variety of applications. Two main types of lasers, CO_2 and Nd:YAG with wave lengths of 10.6 and 1.06 μm, respectively, have been used to investigate the weldability of magnesium alloys. The availability of high-power continuous-wave (CW) carbon dioxide (CO_2) and neodymium-doped yttrium aluminium garnet (Nd:YAG) lasers and the limitations of current welding technology have promoted interest in deep-penetration welding in the past twenty years using these devices. The CO_2 laser has high power output, high efficiency, proven reliability and safety. The weldability of magnesium alloys was reported to be significantly better with the Nd:YAG laser due to its shorter wave length. Leong investigated welding AZ31 alloy using both CO_2 and pulsed Nd:YAG laser, and illustrated the difficulties of CO_2 welding. He concluded that Nd:YAG laser was more suitable for magnesium alloy welding (Leong, 1998).

2.2. Laser power

The penetration depth with laser welding is directly related to the power density of the laser beam and is a function of incident beam power and beam diameter. For a constant beam diameter, penetration typically increases as the beam power is increased. It is well known that high power density at the workpiece is crucial to obtain keyhole welding and to control the formation of welds. The effect of laser power on the penetration depth and weld width for WE43 alloy welded at a speed of 33 mm/s and a focused diameter of 0.25 mm are shown in Fig. 3a and 3b, respectively (Dhahri et al., 2001). Fig. 3 also clearly shows that the threshold power for deep penetration mode welding of cast WE43 alloy is approximately 1kW for a CO_2 laser.

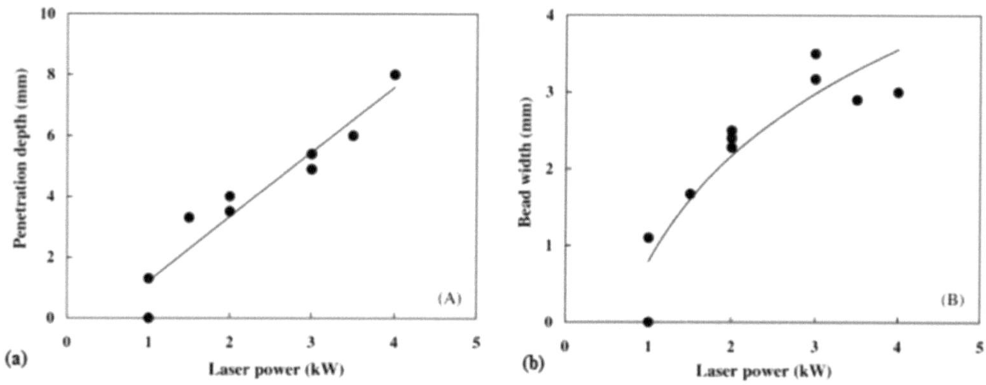

Figure 3. Effect of CO_2 laser power on (a) penetration depth, and (b) bead width of cast WE43 alloy joints (Dhahri et al., 2001).

High beam powers led to deep and wide beads, and reduce both ripples and crowning. Lower irradiance incident upon the workpiece would reduce the spatter as well as the loss of high vapour pressure constituents. It is reported that lower power level and slower speed lead to better weld quality (Cao et al., 2006). A loss of tensile strength was found with Nd:YAG laser power lower than 2kW (Lehner and Reinhart, 1999). For ZE41A-T5 sand castings keyhole welding is reached at a threshold irradiance of 1.5×106 W/cm2 for the machined surface conditions using 2.5 kW Nd:YAG laser power, but the keyhole mode is obtained at 4.0×105 W/cm^2 for the as-cast surface conditions (Cao et al., 2005). It can be concluded that the as-cast surface requires slower power density for the formation of keyholes indicating that the as-cast surfaces have higher energy absorptivity for Nd:YAG laser beams probably due to the coarser surface conditions.

2.3. Welding speed

The effects of welding speed on penetration depth and weld width at different levels of power for CO_2 and Nd:YAG lasers are shown in Fig. 4. It can be seen that the penetration depth and weld width both decrease linearly with increasing welding speed.

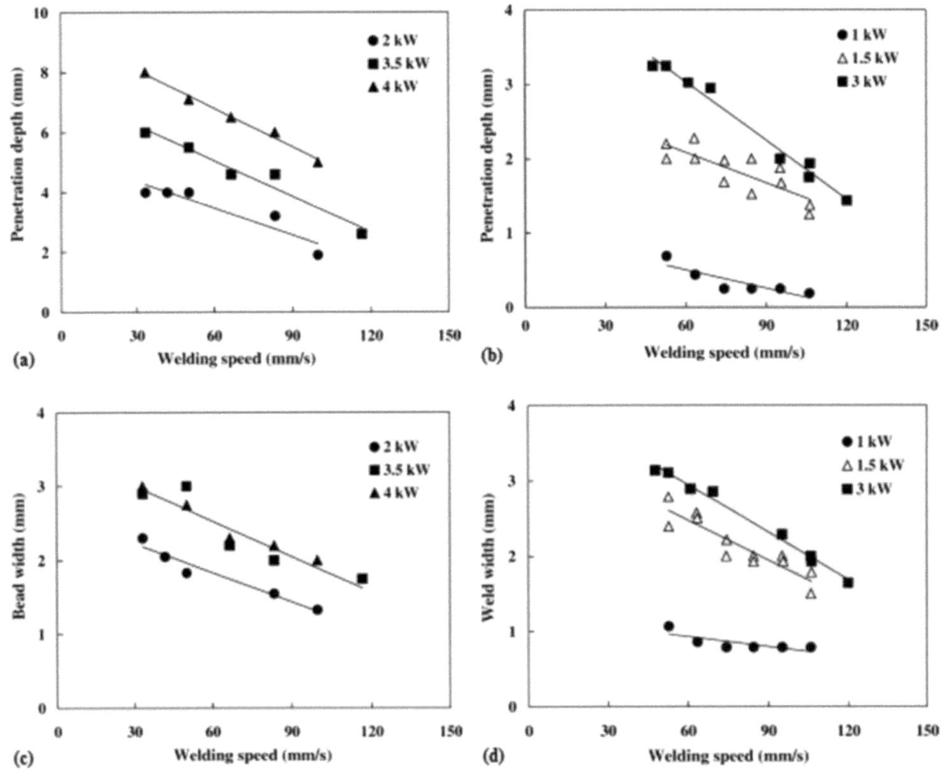

Figure 4. Effect of welding speed on penetration depth for (a) cast WE43 alloy joints welded using a CO_2 laser and (b) die-cast AM60B alloy joints welded using a Nd:YAG laser as well as the effect of welding speed on weld width for (c) cast WE43 alloy joints welded using a CO_2 laser and (d) die-cast AM60B alloy joints welded using a Nd:YAG laser (Marya and Edwards, 2001).

It is well known that a further decrease in welding speed led to little increase in penetration depth but there are increases in weld and HAZ widths (Dhahri et al., 2000). High welding speed was reported to reduce ripples but greatly increase crowning, or even to increase tendency to brittleness in the fusion zone (Cao et al., 2006).

Fig. 5 indicates the effect of welding speed on the penetration depth as well as on the area of molten weld pool for wrought AZ21A and die-cast AZ91D alloys welded using a 1.7 kW CW Nd:YAG laser (Marya and Edwards, 2001). Though similar welding parameters are used, various magnesium alloys exhibit different welding performance due to their different thermo-physical properties.

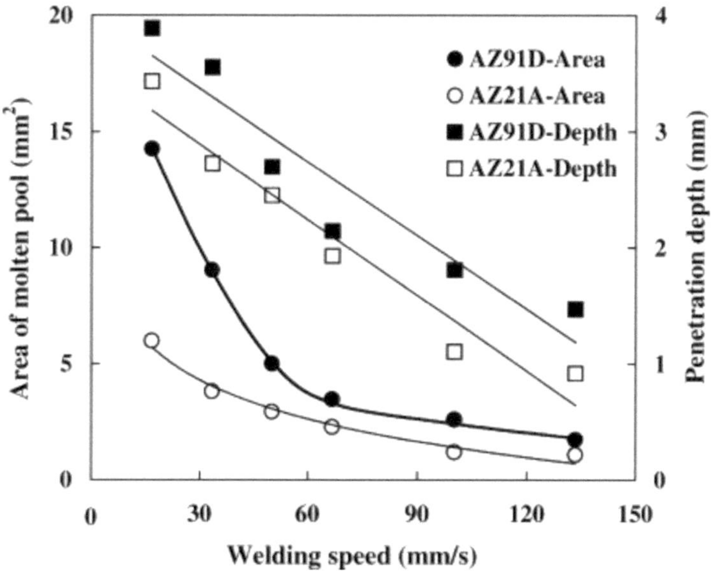

Figure 5. Area of molten weld pool and penetration depth versus welding speed for AZ21A and AZ91D alloys using a 1.7 kW CW Nd:YAG laser (Marya and Edwards, 2001).

Die-cast AZ91D has a lower thermal conductivity of 51 W/mK as compared with 139 W/mK for wrought AZ21A alloy. Thus, the AZ91D alloy has a higher weld depth and weld volume compared with AZ21A alloy. Because of the difference in their thermal properties, much more studies, therefore, is needed to systematically investigate the laser welding characteristics of different magnesium alloys.

When the laser energy input was more than that required for a full penetration, large holes or weld dropout could be observed. Weld drop was observed in both CO_2 and diode conduction welding. This could be due to the long interaction time or the higher power density, while the low viscosity liquid is not able to sustain the welding pool. This defect would significantly reduce the load capability of the weld. Fig. 6 is an example of a CO_2 weld drop. The energy input could be controlled by controlling the welding speed to avoid over-melting, and hence large hole formation and weld dropout (Zhu et al., 2005).

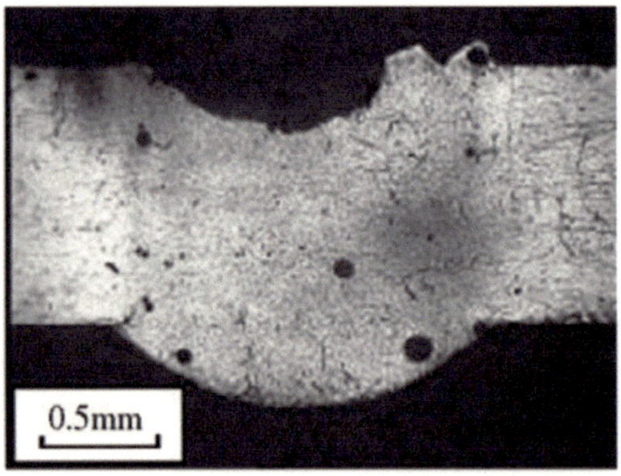

Figure 6. Weld dropout defect in laser welding of AZ31 (Zhu et al., 2005).

It may be proposed whether crowning, ripples, and weld cross-section morphology can be related. It was found high beam powers reduce both ripples and crowning. High welding speeds although reduces ripples, greatly increases crowning. It is clear that these two phenomena are independent. The ripple phenomenon has been associated with growth rate fluctuations due to the effects of surface tension of the weld pool during solidification. Fig. 7 shows the crown increased rapidly with increasing the welding speed and decreasing the beam power. The phenomenon of crowning was found to be primarily dependent upon the welding speed. Large welding speed and low beam powers promote the crowning (Marya and Edwards, 2001).

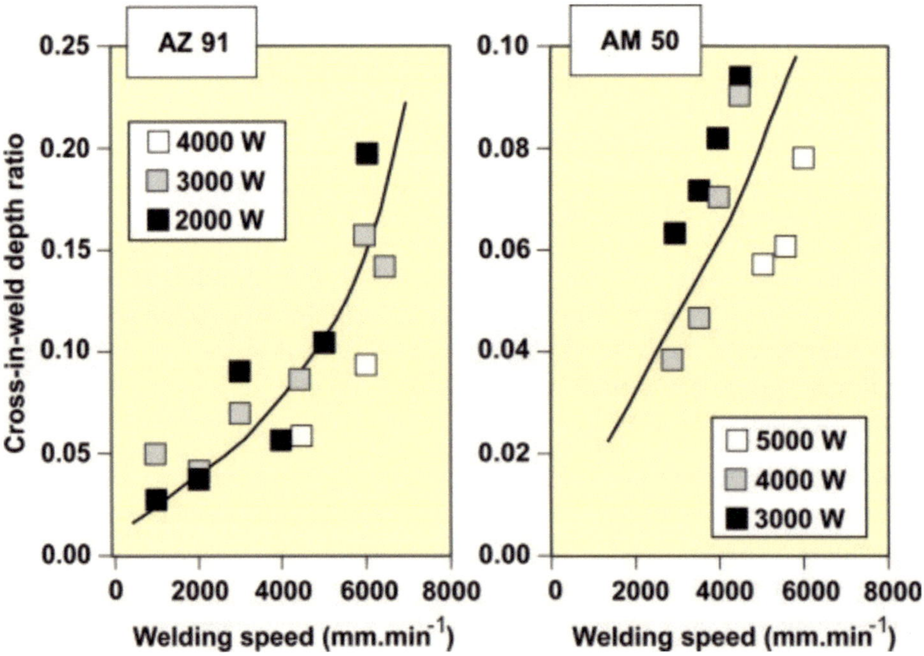

Figure 7. Effects of the beam speed and the beam power on the magnitude of the weld crown relative to the weld depth (Marya and Edwards, 2001).

2.4. Shielding gas

It is well known that magnesium is highly susceptible to oxidation and thereby elaborate protection from the atmosphere is required (Dilthey et al., 1998). This can be achieved using inert gases. Good shielding can prevent burning or porosity and protect the laser optics from metal slag. The shielding gas also influences the formation of the plasma (Dhahri et al., 2001b). Helium, with a high ionization potential of 24.5 eV and good thermal conductivity, has a high plasma formation threshold. Thus, little plasma is produced using helium as shielding gas (Cao et al., 2006).

To identify the best shielding gas, blind welds of AZ91 alloy were produced using helium, argon and nitrogen (Yu et al., 2009). Helium gas is proved to be the best according to the surface roughness, penetration depth, and seam shape factor (depth/width). Studies on welding of WE43 alloy using a 5 kW CO_2 laser reveals that gas flow of He lower than 50 l/min may cause spraying or collapse of the molten pool (Dhahri et al., 2001a). A comparative study of two kinds of inert shielding gases He and Ar for laser welding process showed that He gas possesses a better shielding effect for the laser generated magnesium alloy pool than Ar gas, because of its higher dissociation energy and better conductivity (Dhahri et al., 2001b). For CO_2 laser welding process, the satisfied seams with smooth surface, deep penetration depth, large ratio of weld depth to weld width and high weld strength, could only be produced under the protection environment of much more expensive He gas. In contrast, for the YAG laser welding process, sound seams could be readily obtained under the protective environment of much cheaper Ar gas. It can be concluded that YAG laser is the first choice for the laser welding of magnesium alloys considering weld quality and economic factors (Dhahri et al., 2001b). It was found that the shielding gas configuration could have a significant effect on the weld quality. Welding configurations are shown in Fig. 8a and 8b.

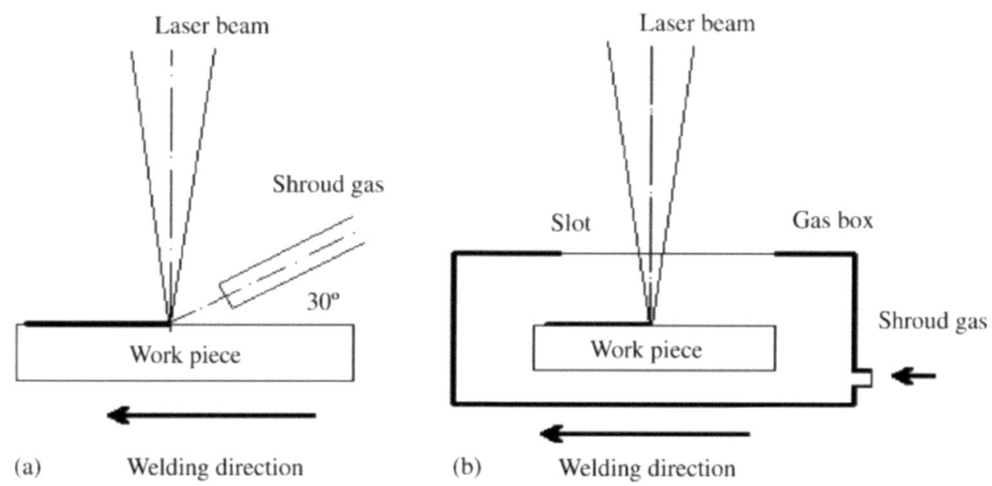

Figure 8. Welding configuration (a) with a shroud gas feed tube and (b) in a gas box (Zhu et al., 2005).

If the gas flow was too weak, there would be insufficient weld protection, while too strong the gas flow would disturb the weld geometry, or cause defects. When the gas from the feed tube above the weld is not aligned with the traverse direction, a side cut would occur as shown in Fig. 9. Besides, if it is in the opposite direction with that, shown in Fig. 2, an irregular weld bead would occur. When the laser welding was conducted in the gas box, a reasonable weld can be made more easily.

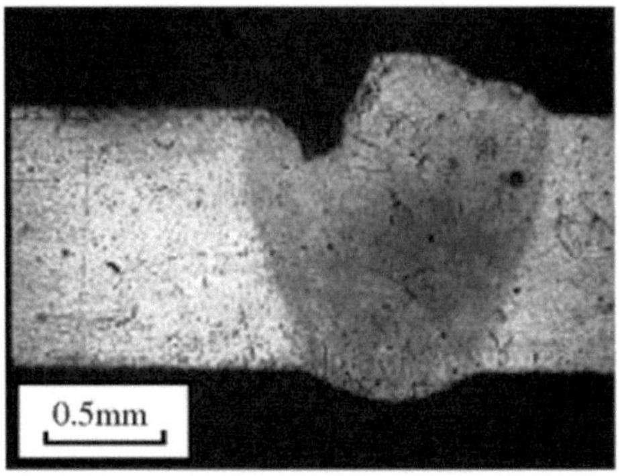

Figure 9. Slide blow defect in AZ31 laser welds (Zhu et al., 2005).

Therefore, gas shroud configuration is very important, which affects the weld zone oxidation and weld beam geometry. The use of an atmospherically controlled glove box is more effective.

2.5. Fiber laser

Fiber laser, as the third generation of industrial laser, is rapidly developing in recent years due to a number of advantages such as compact size, long transmission distance, high efficiency, superior beam quality and high power output (peak output is up to 50 kW). In the fields of joining steel and aluminium alloys, fiber laser has made a big progress in experimental studies and industrial applications, showing a great potential to replace conventional Nd:YAG and CO_2 lasers. However, few studies on welding magnesium alloys by fiber lasers are conducted. It is reported that the UTS of the joint decreases, but the yield strength increases after fiber laser welding of fine-grained magnesium alloy (Yu et al., 2009). An IPG YLR 6000W fiber laser and a KUKA KR60 robot are shown in Fig. 10.

Using of 6kW fiber laser to weld AZ31B wrought magnesium alloy showed that the impact of welding speed on the weld penetration depth is less than that of laser power, though the weld penetration depth increases with decreasing the welding speed (Wang et al., 2011). The effects of welding parameters on weld penetration depth are depicted in Fig. 11. It can be found that the weld penetration depth only increases by 0.5 mm or so as the welding speed decreases from 4 m/min to 2 m/min when the laser power is kept the same. On the other hand, laser power plays a big role in the characterization of fiber laser welding of AZ31B magnesium alloy. The weld penetration depth increases by 2–3 mm per kilowatt in laser power. It is found that in the range of 2.5 to 4.0 kW laser power, stable welding process and accepted welds without macro-defects can be obtained (Wang et al., 2011).

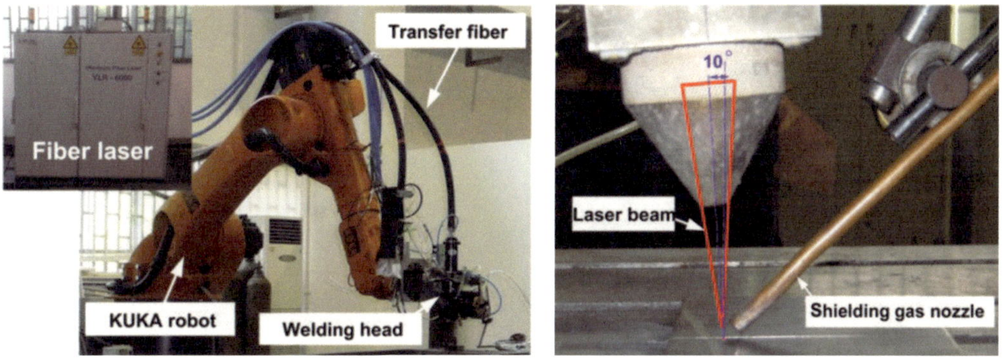

Figure 10. Diagrams of fiber laser welding system and experimental arrangement, (a) IPG fiber laser installed on a KUKA robot, (b) welding head (Wang et al., 2011).

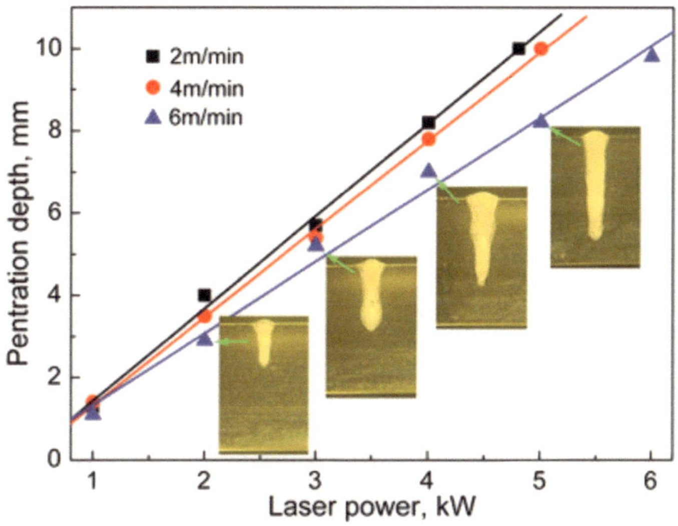

Figure 11. Effects of welding parameters on weld penetration depth (Wang et al., 2011).

2.6. Microstructure

In the laser welding, three distinct regions can be distinguished in the weld cross section, i.e., the fusion zone (FZ), partially melted zone (PMZ), and the heated affected zone (HAZ), which are categorized according to the temperatures experienced during welding. A narrow weld joint is an important characteristic of high power density welding. FZ, PMZ and HAZ for ZE41A-T5 alloy joint are shown in Fig. 12.

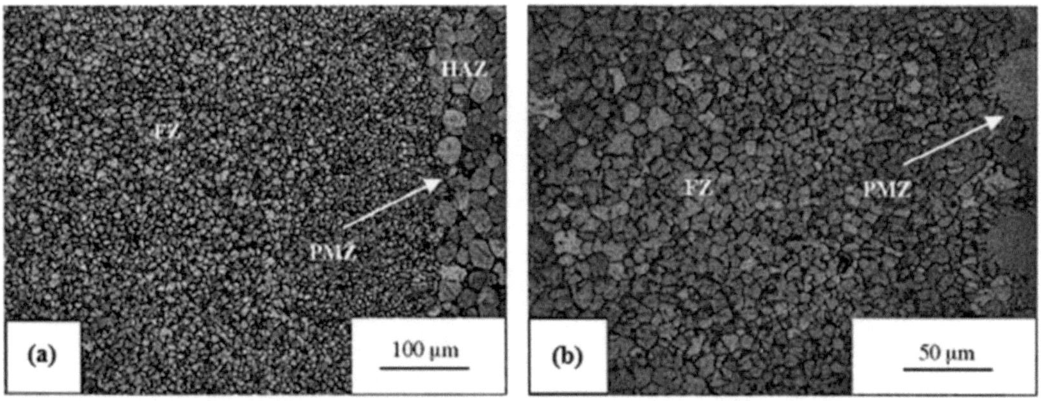

Figure 12. Optical micrographs showing (a) microstructure near the interface between the FZ and HAZ for a ZE41A-T5 alloy joint and (b) close-up view of Fig. 12a (Cao et al., 2005).

The 2.5 kW CW Nd:YAG laser welding of 2 mm ZE41A-T5 sand castings showed that the fusion zones have widths of approximately 0.8–1.3 mm and the width of HAZ is approximately 2 mm and the partially melted zone is rather narrow, only several grains wide. It is showed that the HAZ in welded AZ91D die castings using a 2 kW CW Nd:YAG laser had a width of 50–160 μm which is depending on welding speed (Haferkamp et al., 1996). Moreover, 2 kWPW Nd:YAG and 6 kW CW CO_2 laser welding of wrought AZ31B alloy indicated that the width of the HAZ was 50–60 μm (Haferkamp et al., 2000). If the base metal is work hardened, the strengthening effect will be removed. If the alloy is in a fully tempered state, the loss of precipitates may lead to the HAZ overaged. If the alloy is susceptible to grain growth, the grains will grow in the HAZ which leads to decrease in mechanical properties (Cao et al., 2006). A significant grain coarsening can be observed in the HAZ in wrought AZ31 alloy, however, in all laser welded cast alloys, no grain coarsening occurs within the HAZ zone. Certainly, grain growth can be minimized at high welding speeds (Cao et al., 2006).

2.7. Metallurgical defects

2.7.1. Porosity

The presence of porosity in the weld is one of the major concerns during laser beam welding of magnesium alloys. The high porosity is also one of the most important factors to reduce the failure strength of welded joints undergoing higher heat input, even though their grain size in the FZ is similar to or even smaller than that of the base metal (Wang et al., 2011). During the tensile test, because of stress concentration around the porosities, the microcrakes usually initiate from these weakest positions, and then propagate to fracture the joints (Zhao and Debroy, 2001).

Magnesium has significantly high solubility for hydrogen in the liquid phase and the porosity in magnesium alloys is dependent on the amount of dissolved hydrogen (Stolbov et al.,

1990). The tolerable hydrogen gas content in welds may depend on a number of factors such as the parameters of the welding process, alloy composition, local solidification time, thermal gradient, weld structure, and inclusion concentration (Cao et al., 2006). However, no acceptable hydrogen content limits have ever been reported for laser welded joints of magnesium alloys. With magnesium alloys containing zirconium, hydrogen will react with zirconium to form ZrH2, and finally in this case, hydrogen porosity will not be a problem (Cao et al., 2006), (Cao et al., 2005). It is found that the rejection of hydrogen from the Mg17Al12 intermetallic compound helped in the nucleation and/or growth of micro-porosity during the last stages of solidification of AZ91 alloy. It was recognized that the porosity in fiber laser welds of AM60, AE42 and AS41 cast magnesium alloys was mainly caused by the growth of the initial pre-existing pores. Imperfect collapse of the keyhole and turbulent flow in the weld pool could also been linked with porosity formation. In the fiber laser of AZ31B wrought magnesium it is proposed that, the initial preexisting micro-pores plays a major role in the pore formation when the heat input increases to 85.7 J/mm or higher, while the collapse of the unstable keyhole is the main reason for the pore formation when the heat input is lower than 62.5 J/mm (Wang et al., 2011). Thus, decreasing heat input is effective to reduce the porosity in fiber laser welds of magnesium alloys. Fig. 13 depicts the effect of welding speed on the pore fraction in AZ91 alloy.

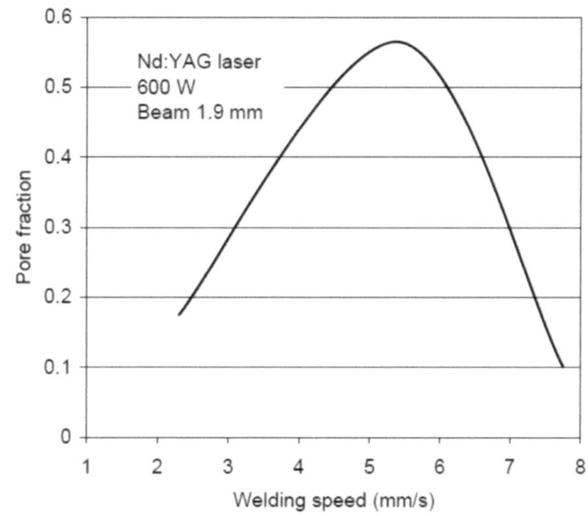

Figure 13. Effect of welding speed on the pore fraction in AZ91 alloy.

The existence of gas inclusions in the base metal is another important factor in contributing to the formation of large pores during Nd:YAG laser welding of magnesium alloys (Zhao and Debroy, 2001). As shown in Fig. 14a hydrogen porosity is the dominant pore in laser welded die-cast magnesium alloys because of their extremely high initial gas contents. Clearly, production of weldable and heat treatable die-castings necessitates the reduction of gas contents in magnesium alloys (Cao et al., 2006). Vacuum die-castings have relatively low initial gas contents.

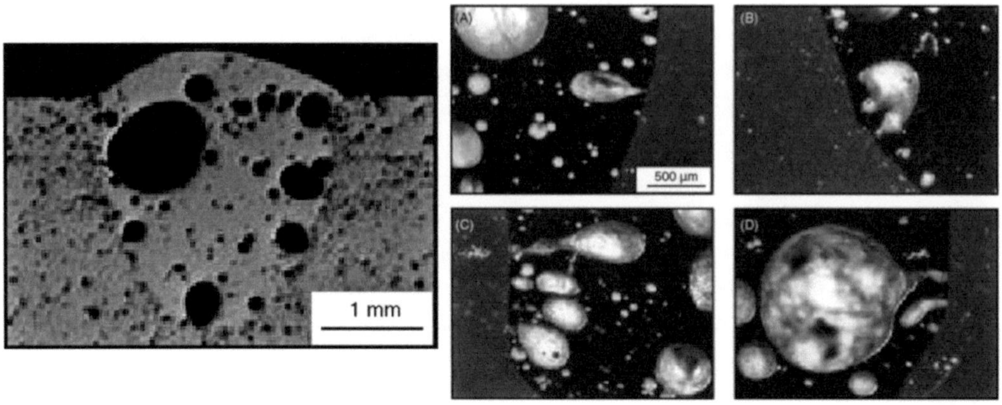

(a) (b)

Figure 14. (a) Pores in die-cast AM60B alloy welded using a 1.5 kW CW Nd:YAG laser (Zhao and Debroy, 2001), and (b) Formation of large pores in the FZ due to the expansion and coalescence of the preexisting pores in base die-cast AM60B alloy (Cao et al., 2006).

The porosity in the FZ increases with the increase in heat input, i.e., increase in the laser power and decrease in the welding speed. It is proposed that well-controlled remelting of the FZ leads to removal of gas bubbles and reduced porosity in the FZ. When die cast magnesium alloys remelt during laser beam welding, the gas can be enlarge due to heating (thermal expansion) and thus leading to the release of the gas pressure. Finally, the growth of porosity leads to expansion and coalescence of the pores, causing the formation of large pores in the fusion zone as shown in Fig. 14b (Cao et al., 2006).

2.7.2. Weld cracking

Hot cracks have been one of the main welding defects for magnesium alloys. In most magnesium alloys, an increase in alloying elements will generally increase the solidification temperature range. The large freezing temperature range, large solidification shrinkage, high coefficient of thermal expansion, and low melting point intermetallic constituents potentially make magnesium alloys susceptible to heat affected zone liquation cracking and solidification cracking in FZs (Marya and Edwards, 2002). In addition, a number of magnesium alloys are thought to be susceptible to stress corrosion cracking. Thus, the welded joints should be used after stress relieving (Cao et al., 2006).

2.7.3. Solidification cracking

Solidification cracking usually occurs in the alloys with large solidification interval, such as Mg–Zn–Zr, Mg–Al–Zn, etc. (Cao et al., 2006). The susceptibility to solidification cracking is evaluated using the circular-patch-weldability test and operating windows of welding parameters are identified (Lathabai et al., 2003). The alloys containing up to 6% Al and up to 1% Zn possess good weldability, while the alloys containing over 6% Al and up to 1% Zn are moderately crack sensitive, because of the occurrence of low melting-point constituents

($Mg_{17}Al_{12}$). Even without Al, the alloys containing over 3% Zn are highly susceptible to solidification cracking and difficult to weld (Marya et al., 2000). On the other hand, hot tearing is mainly influenced by joint composition, microstructure, welding process parameters, and joint design. More investigation is needed to prevent the hot cracking.

2.7.4. Oxide inclusions

It is well known that, oxides are the main inclusions in magnesium alloys. It is found that, the presence of oxide inclusions in the base metal is thought to be an important factor in the formation of large pores during Nd:YAG laser welding of magnesium alloys (Haferkamp et al., 1996). Oxidation of magnesium is increased at high temperatures, resulting in the formation of a surface magnesium oxide layer (Cao et al., 2006). The amorphous structure of MgO films in solid pure magnesium generally performs an important role as a protective layer to prevent further oxidation at low temperature, but the films easily become porous and loose with the temperature increase. It is reported that the addition of Ca to the pure Mg as well as the addition of Al and Y to the Mg–Ca alloys can further improve oxidation resistance. The entrapment of surface oxides into the molten pool during welding is an important source for oxide inclusions in laser welded magnesium alloys. Finally, it can be proposed that the surface oxides are formed during three stages: (i) oxygen chemisorptions, (ii) formation of the oxide layer (nucleation and lateral growth) and (iii) oxide thickening (Berlin et al., 2011). On the other hand the existing oxides in the base metal originating during primary metal processing or during casting is another source.

2.7.5. Loss of alloying elements

The relatively high vapour pressure of liquid magnesium compared with aluminium points to the potential problem of evaporative losses, particularly if zinc is also present as an alloying addition. The evaporation will cause a variation of chemical composition in the FZ, especially at high laser power density. Minimizing the irradiance incident upon the workpiece would decrease the loss of high vapour pressure elements (ASM, 1993). It is reported that low welding speed leads to larger reductions of both Mg and Zn. it is expected that evaporative loss is more problematic in low zinc-zirconium alloys where zinc provides strengthening and zinc evaporation might decrease the properties to unacceptable levels. Further work is needed to fully understand the main influencing factors to build up the quantitative relationship between the element loss of evaporation and welding process parameters.

3. Gas tungsten arc welding

Gas-Tungsten Arc Welding (GTAW), also known as HeliArc, Tungsten Inert Gas (TIG), and tungsten arc welding, was developed in the late 1930s when a need to weld magnesium became apparent. Russell Meredith developed a welding process using the inert gas helium and a tungsten electrode to fuse magnesium. TIG welding is used extensively for welding stainless steel, aluminium, magnesium, copper, and reactive materials (for example, titanium and tantalum). This joining method replaced riveting as a method of building aircraft with aluminium and magnesium components (ASM, 1993). This process uses the heat, generated by an electric arc struck between a non-consumable tungsten electrode and the workpiece, to fuse metal in the joint area, and produce a molten weld pool. The arc area is shrouded in an inert or reducing gas shield to protect the weld pool and the non-consuming electrode. The process may be operated autogenously (without filler), or filler may be added by feeding a consumable wire or rod into the established weld pool. Fig. 15 illustrates the schematic of TIG welding process.

Advantages and limitations of GTAW include (ASM, 1993):

- Produces high quality, low distortion welds
- Free of the spatter associated with other methods
- Can be used with or without filler wire
- Can be used with a range of power supplies
- Welds almost all metals, including dissimilar ones
- Gives precise control of welding heat
- Produces lower deposition rates than consumable electrode arc welding processes
- Requires slightly more dexterity and welder coordination than gas metal arc welding or shielded metal arc welding for manual welding
- Less economical than consumable electrode arc welding for thick section greater than 9.5 mm (3/8 in)
- Problematic in blowy environments because of difficulty in properly shielding the weld zone

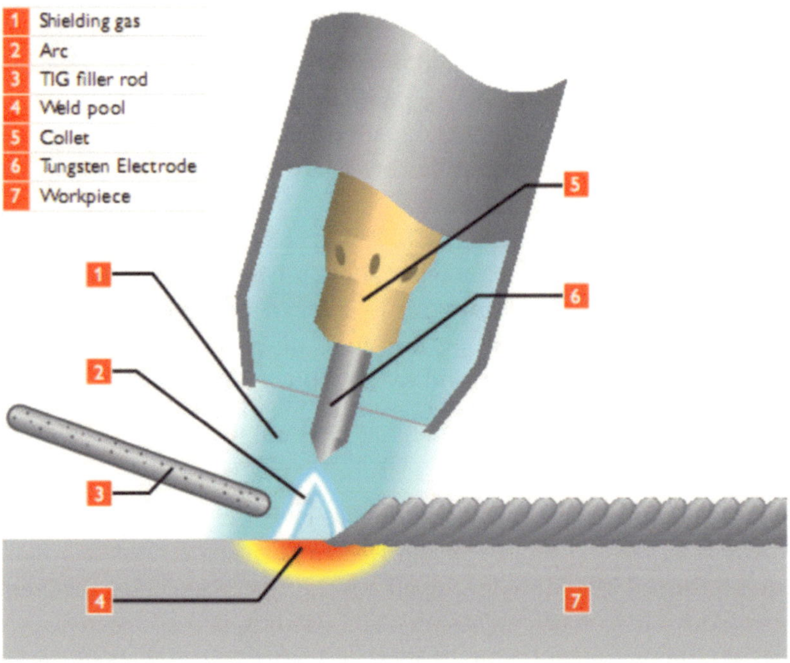

Figure 15. Schematic of TIG welding process (a) overall process; (b) welding area enlarged.

Direct or alternating current power sources with constant current output characteristics are normally employed to supply the welding current. For direct current (DC) operation, the tungsten may be connected to either output terminal, but is most often connected to the negative pole. Shielding gas is directed into the arc area by the welding torch, and a gas lens within the torch distributes the shielding gas evenly over the weld area. In the torch, the welding current is transferred to the tungsten electrode from the copper conductor.

3.1. Welding current

Current is one of the most important process parameters to control in any welding operation, because it affects the penetration depth, welding speed, deposition rate, and weld quality. The TIG process may be operated in one of the following modes:

Direct Current Electrode Negative (DCEN): in this mode the tungsten electrode is the negative pole in the welding circuit, the workpiece being the positive pole (fig. 16a).

Direct Current Electrode Positive (DCEP): in this mode the tungsten electrode is the positive pole in the welding circuit, the workpiece being the negative pole (Fig. 16b).

Alternating Current (AC): in this mode the polarity of the tungsten electrode and the workpiece alternate between negative and positive at the frequency of the applied welding current.

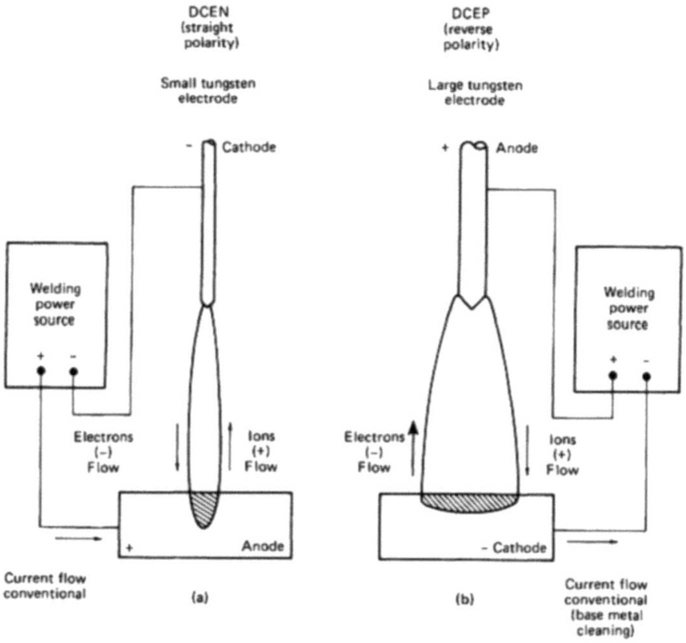

Figure 16. Effect of polarity on GTAW weld configuration when using direct current. (a) DCEN, deep penetration, narrow melted area, approximate 30% heat in electrode and 70% in base metal. (b) DCEP, shallow penetration, wide melted area, approximate 70% heat in electrode and 30% in base metal (ASM, 1993).

In TIG welding of light metals, the alternate current is used, because the rapidly changing polarity gives combination of both cathodic cleaning action, which is beneficial for oxide removal during welding of aluminium and magnesium, and lower heat input. It is found that welding current affect the weld shape and arc voltage. The effect of welding current on weld penetration (D) and weld depth/width (D/W) ratio as well as arc voltage in AZ31B magnesium alloy with and without $CdCl_2$ flux is shown in Fig. 17. It can be demonstrated that, with increasing welding current, the weld D, the weld D/W and arc voltage all increasing. The large welding current will increase the electro-magnetic force, which strengthens the downward body convection in the welding pool, leading to an increase in the D/W ratio.

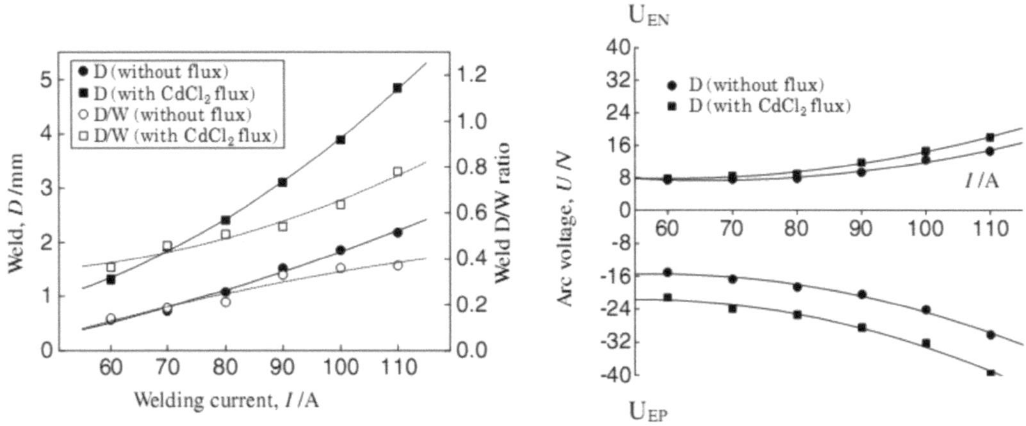

Figure 17. Effect of welding current on weld D, weld D/W ratio and arc voltage with and without $CdCl_2$ flux (Marya and Edwards, 2002).

3.2. Shielding gases

The preferred gas for the AC-TIG welding of magnesium is argon, although helium and argon–helium mixtures may be used. Argon gives a wide, shallow penetration weld bead, but will leave the weld bright and silvery in appearance. Argon has a low ionization potential (2.52×10^{-18} J, or 15.7 eV), making it easier to form an arc plasma than with other shielding gases. The easiest arc ignition and most stable arc will also be achieved with argon. Argon, because of its high density, must be used at lower flow rates than helium. Typical flow rates for argon are 7 L/min (15 ft^3/h), and for helium, 14 L/min (30 ft^3/h).

For transition metal alloys, when the welding path is coated with chemical fluxes, a technique exists of flux-assisted gas tungsten arc welding (FA-TIG). Generally, fluxes allow full penetration welding at greater rates using relatively inexpensive gas tungsten arc as the heat source. Using argon shielding and chloride fluxes, magnesium alloy welding tests showed an increase in weld penetration (Marya and Edward, 2002). Among several chlorides, including LiCl, $CaCl_2$, $CdCl_2$, $PbCl_2$ and $CeCl_3$, cadmium chloride is the most effective. The weld penetration of the GTA weld prepared with a CdCl2 flux is twice of that for the weld prepared without flux (Zhang et al., 2008).

4. Resistance spot welding

Resistance spot welding is the localized joining of two or more metal parts together by the heat, generated using resistance to the flow of electric current through workpieces, which are held together under force by electrodes. The contacting surfaces in the region of current concentration are heated by a short-time pulse of low-voltage and high-amperage current. As a result, at the interface of welding sheets a molten pool (weld nugget) is formed. When the flow of current ceases, the electrode force is maintained till the weld metal cools down and solidifies. The electrodes are retracted after each weld, which usually is completed in a fraction of a second. RSW can be separated into Silicon Controlled Rectifier (SCR) AC resistance welding and inverter DC resistance welding (Hwang et al., 2010). For high electrical conductivity of magnesium and low heat generation in the weld, high welding currents are required. The process is used mainly in assembly lines to weld products made of thin gauge metals, and it has potentials for joining sheets of magnesium alloys (Xiaoa et al., 2011).

4.1. Weld microstructure

The joint, achieved by a technique of resistance spot welding, consists of the weld nugget and the heat affected zone. The weld nugget consists of the cellular dendritic structure at nugget edge, accompanied by equiaxed dendritic structure within its centre for AZ31 alloy (Fig. 18). The diameter of equiaxed dendrites in in the weld nugget of plates with lower thickness is smaller than that in the plates with higher thickness, and the length of columnar dendrites in the vicinity of the fusion boundary is also shorter.

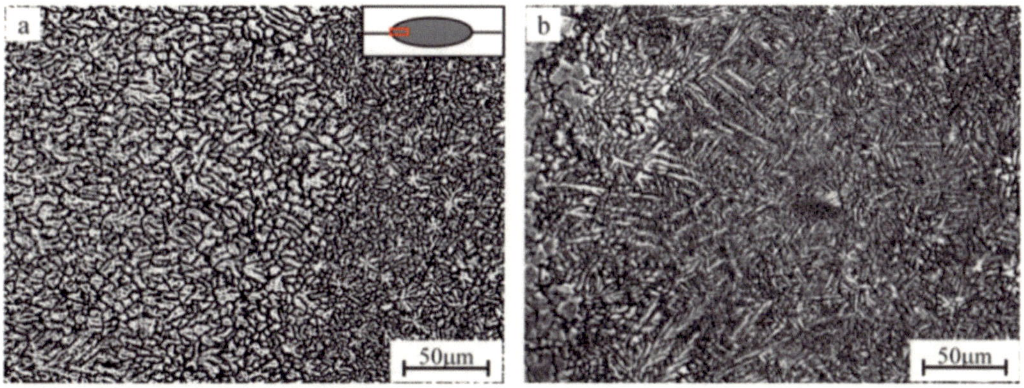

Figure 18. Equiaxed dendritic structure in the fusion zone of AZ31 Mg alloy resistance spot welds: (a) plate with a thickness of 2 mm and (b) plate with a thickness of 1.5 mm (Xiaoa et al., 2011).

The refined fusion-zone microstructure results in a longer fatigue life than welds with a coarse fusion-zone microstructure, when interfacial failure across fusion zone occurs at a higher level of cyclic load range (Xiaoa et al., 2011). The boundary melting and coarsening was observed in the heat affected zone. It is proposed that the weld nuggets have a tendency to hot cracking. The joint strength and cracking susceptibility are influenced by the welding current. While the higher current increases strength it also increases the nugget tendency to crack.

4.2. Effect of welding parameters on joint properties

As weld current increases, the nugget width increases as well. This effect is proved for both SCR AC resistance welding and inverter DC resistance welding which is shown in Table 1. In both RSW welding technics cracks and pores have been found, affecting the properties of weld. In the case of magnesium alloys, the welding part rapidly expands and retracts during welding and quickly cooling down, because of high coefficient of linear expansion. Then, a blank in the expanded space becomes pore (Yin et al., 2010). Thus, it is necessary to perform more investigations to reduce pore and crack.

200 kgf/cm², 9 cycle

Current (kA)	13	15	17	19
SCRAC	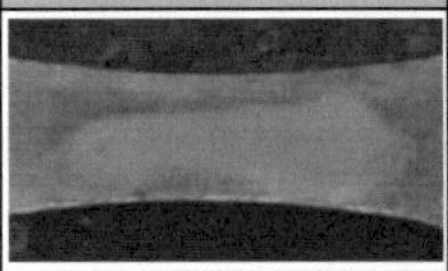	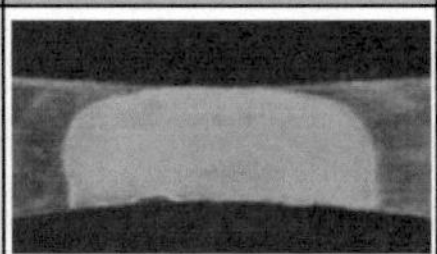	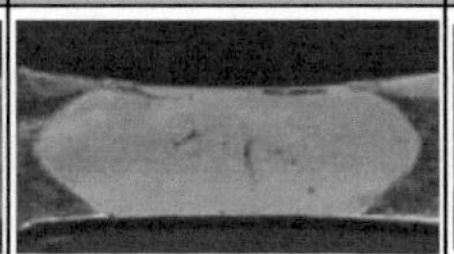	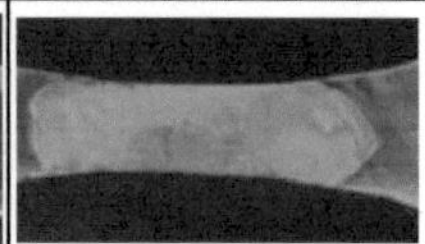
Nugget (mm)	3.5	4.0	4.6	5.8
Inverter DC	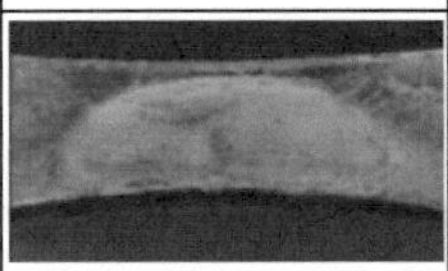	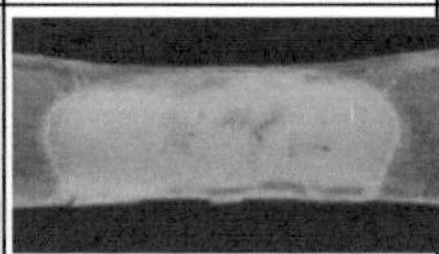	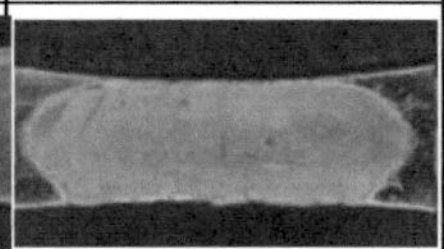	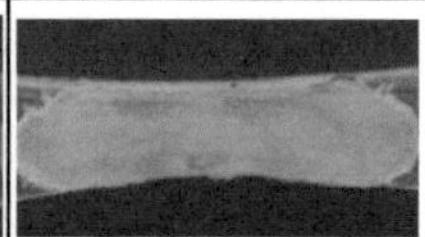
Nugget (mm)	3.9	4.4	5.0	6.0

Table 1. Effect of welding current on nugget shape and size (Hwang et al., 2011).

It is confirmed that, the inverter DC resistance welding is more efficient than SCR AC resistance welding, in terms of heat input and weldability through shear tension strength, macro section and welding lobe (Hwang et al., 2011).

Fig. 19 shows the effect of selected process parameters on joint properties. With increasing the current, weld integrity reduces at the constant electrode force, as seen through porosity level (Fig. 19a). An increase in the electrode force reduces the role of current, and at sufficiently high electrode force the effect of current disappears. In the case without steel plates, the reduction in pore formation was achieved by increasing the electrode force and extending the holding time after current shut-off. The relative importance of holding time after current shut-off depended on welding current.

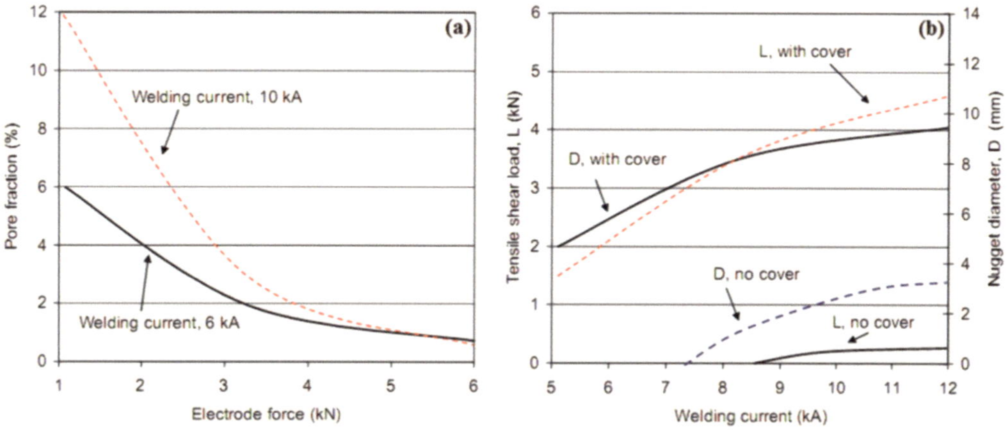

Figure 19. Resistance spot welding of AZ31B alloy with and without steel cover plates: (a) pore fraction versus electrode force; (b) tensile shear load and nugget diameter as a function of welding current (Shi et al., 2010).

5. Electron beam welding

Electron beam welding (EBW) is a process that melts and joins metals, employing a dense stream of high velocity electrons to bombard, heat, and melt the materials being joined. The electron beam is generated by electron gun, composed of a tungsten cathode and an anode placed in high-vacuum. Fig. 20 illustrates the schematic of EBW process. The beam currents and the accelerating voltages employed for typical EBW vary over the ranges of 50–1000 mA and 30–175 kV, respectively. A high intense electron beam can vaporize the metal and form a vapour hole during welding that is a keyhole (Fig. 20b). EBW process is suitable for difficult welding, where high repeatability is needed. Although laser welding is a very convenient method, its energy limitations restrict the degree of fusion penetration in thicker workpieces (Powers and Laflamme, 1992). Five factors control the EBW process and affect the quality of the weld, including accelerating voltage (V, kV), beam current (I, mA), welding speed (υ, mm/s), vacuum level (below10^{-3} Torr), and the spot size (φ, mm) of the electron beam on the surface of the workpiece (Chi et al., 2006).

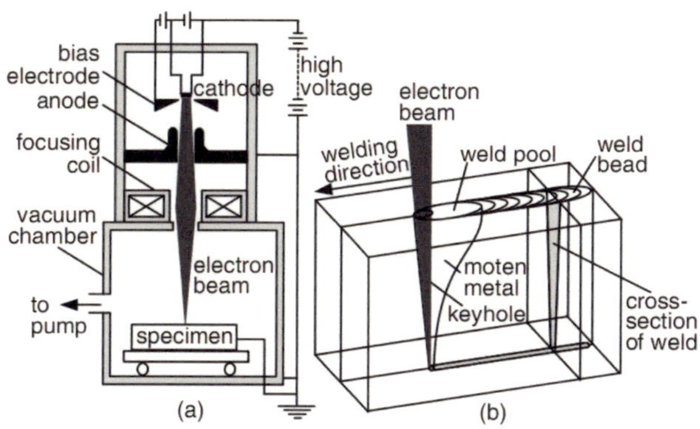

Figure 20. Electron beam welding: (a) process; (b) keyhole (Kou, 2003).

Advantages and limitations of RSW include (ASM, 1993):

- With a very high power density in EBW, full-penetration keyholing is possible even in thick workpieces.
- The total heat input per unit length of the weld is much lower than that in the arc welding, resulting in a very narrow HAZ and little distortion.
- Dissimilar metals can be welded because of the very rapid cooling which can prevent the formation of coarse, brittle intermetallic compounds.
- Equipment costs for EBW generally are higher than those for conventional welding processes. The requirement of high vacuum (10^{-3}–10^{-6} torr) and x-ray shielding is inconvenient and time consuming.
- The available vacuum chamber capacities and workpiece size are limited. The fine beam size requires precise fit-up of the joint and alignment of the joint with the gun.

In the case of thermal effect during welding of AZ61 alloy, it is proposed that the welding heat source of electron beam produces two special thermal effects: (i) deep-penetration thermal effect and (ii) surface thermal effect of metal vapour (Luo et al., 2010). The experimental data for AZ61 alloy show that the key parameters affecting the keyhole thermal effect are the welding heat input and focus coil current, which has also an influence on the weld shape.

Increasing the Al content in EBW of AZ series of magnesium alloys (up to10%) leads to increasing both the strength and micro-hardness of the weld, and decreases its ductility, because of the increase in concentration of brittle precipitates in the FZ (Chi et al., 2006).

During electron beam welding of 11 mm thick AZ31B plates with a power of 4000 W to 5000 W the effect of various process parameters is examined (Chi and Chao, 2007). The factors reducing the weld strength are: deviations of weld geometry, porosity and grain coarsening. In general, the weld strength reaches over 90% of that for the base alloy.

6. Electromagnetic welding

Electromagnetic welding is a phenomenon in which the current carrying conductors exert a force on each other. Fig. 21 shows a schematic diagram of magnetic pulse welding (MPW). This welding process can be classified as an impact welding technique, which is generally a set of processes that use a rapid energy source to accelerate and cause collision between two metal surfaces. The collision must take place at a high enough velocity to achieve bonding through contaminant and oxide removal. The sheet surfaces must make contact at a small angle, typically at least 4°, up to ~30°. This creates a single dynamic collision point that travels along the surfaces as they collide and bend flat against one another. Some advantages of MPW process could be (Berlin, 2011): base-metal like bond strength, extremely high speed, dissimilar metal combination, no filler or shielding gas, and no HAZ.

Fig. 22 depicts β (collision angle) and schematically represents the deformation taking place during double sided MPW, as carried out in AZ31 magnesium alloys. The joint morphology consisted of three distinct regions observed in cross section: an unbounded centre zone ~3 mm wide, twin bond zones ~1 mm wide, and the outer unbounded surfaces which span the rest of the impact flattened area. The initial contact between the sheets was normal and the angle β is negligibly small. Each bulge acts as a rigid barrier to the others forward motion. Two collision fronts progress symmetrically outwards from the initial contact area of the bulge peaks. Each of the twin bonds have the exact opposite welding direction of the other as well as β at each front increases continuously (Berlin, 2011). The location between two bounded zones called unbounded centre zone.

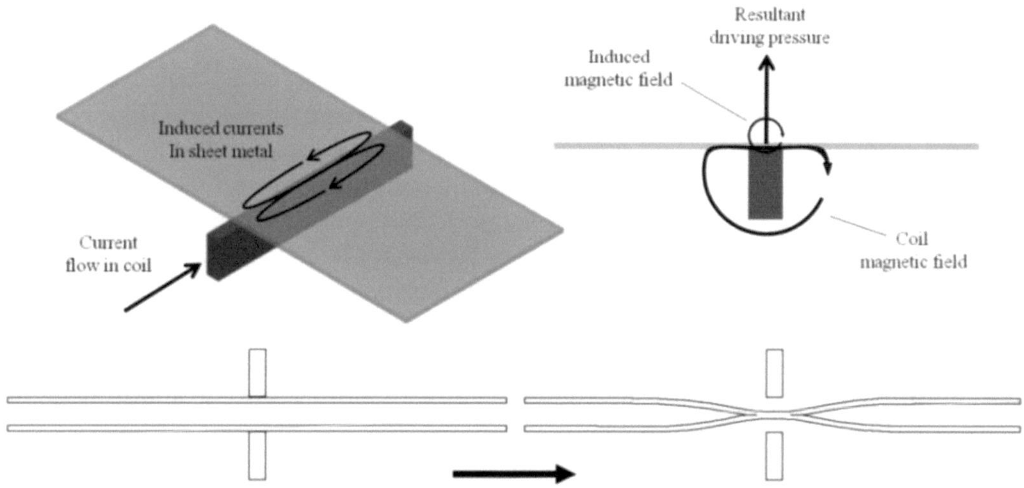

Figure 21. Schematic of the magnetic pulse welding process (Berlin, 2011).

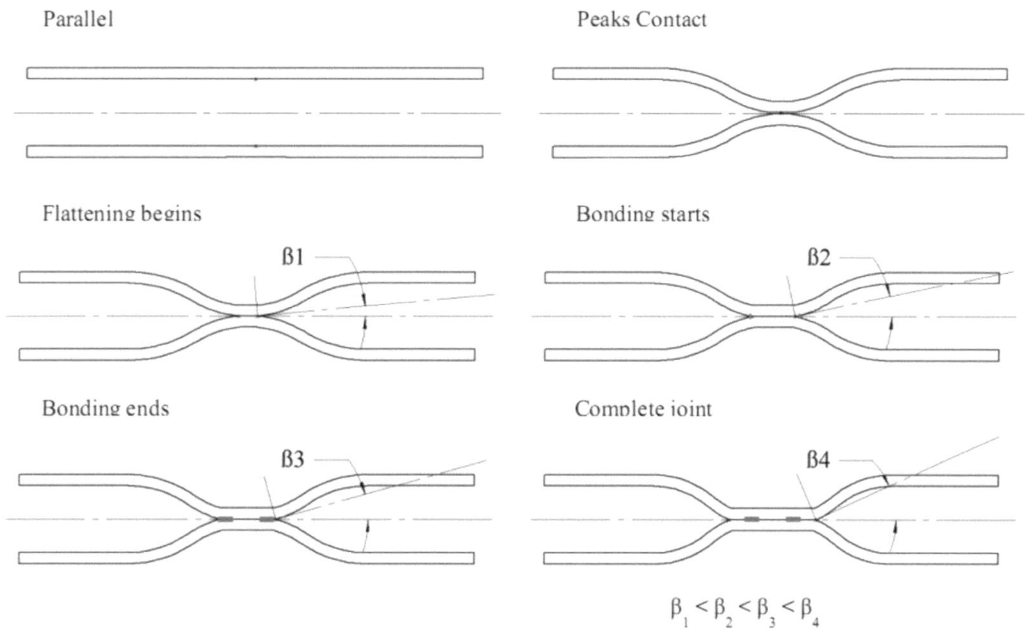

Figure 22. Schematic diagram showing the initial impact and flattening of the sheets during MPW (Berlin, 2011).

Aizawa was the first to successfully weld two flat Al sheets in a lap joint configuration using MPW (Aizawa, 2003). Recently, most MPW researches are focused on joining dissimilar metals, with primary interest in the joint mechanical properties (Berlin et al., 2011). At present, there are some systematic investigations on the joining of magnesium to magnesium alloys using MPW or other forms of impact welding. During lap joints of magnesium AZ31 sheets using magnetic pulse welding, it is found that the bond displays high shear strength, almost equivalent to the base metal shear strength (Aizawa, 2003).

MPW is a technique provided satisfactory results for Al-AZ91 joint. The sheets with a thickness of 0.5 to 1 mm were seam welded, producing a weld having the width of 5 mm. The weld thickness was 10% less than the original sheet thickness. The weld zone was created as a result of combined effect of heating by eddy currents and the strong impulse electromagnetic force (magnetic pressure) and no obvious fusion boundary was microscopically detected in the joint interface. The difference in electrical conductivities of Al and Mg led to the skin depth of 0.7 mm in Al and 0.6 mm in Mg alloys. Additionally, for the experimental conditions, the minimum discharge energy required for Electromagnetic welding process of AZ31 to AA 3003 sheet was found to be in range of 4.3-5.5 kJ (Kore et al., 2009). Further increase in energy results in severe plastic deformation and failure away from the weld in the plastically deformed zone.

The electromagnetic compression of tubular profiles with high electrical conductivity is an innovative joining process for lightweight materials. The components are joined using pulsed magnetic fields which apply radial pressures of up to 200 MPa to the material, causing a symmetric reduction of the diameter with typical strain of about 4^{-10} s^{-1}. Since there is no contact between components to be joined and the joining machine, there is not possible damage of the welded parts. The method was examined for aluminium alloys and has potentials for magnesium.

7. Diffusion welding

Diffusion welding (also known as diffusion bonding) is a metal joining process that requires the application of controlled pressures at elevated temperatures and usually a protective atmosphere to prevent oxidation (ASM, 1993). No melting and only limited macroscopic deformation or relative motion between the joining surfaces of the parts occurs during bonding. Since temperature does not exceed melting, it allows to eliminate many problems associated with fusion welding.

The mechanism of bond formation in diffusion welding is believed to be the deformation of the surface roughness in order to create metal-to-metal contact at asperities, followed by the removal of interfacial voids and cracks by diffusional and creep processes (ASM, 1993). There are three major stages of the bonding progress (Czerwinski, 2010) which are illustrated in Fig. 23. First, a contact between materials occurs through the mating surfaces. During the second stage, diffusion within grain boundaries predominates, thus eliminating pores and ensuring arrangements of grain boundaries. During the third stage, the volume diffusion dominates and process is completed.

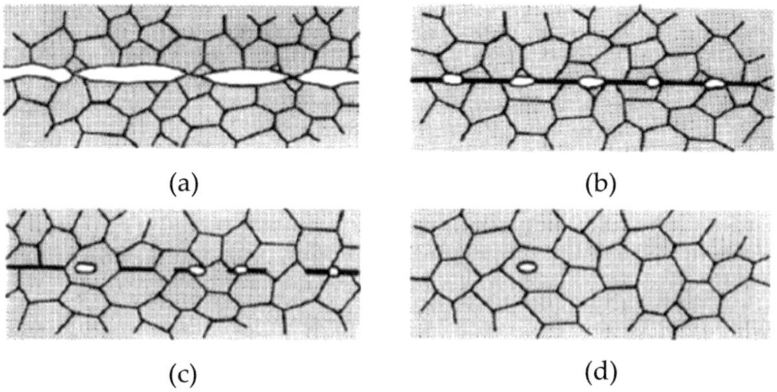

Figure 23. Three-stage mechanistic model of diffusion welding. (a) Initial asperity contact, (b) first stage deformation and interfacial boundary formation, (c) Second stage grain boundary migration and pore elimination, (d) Third stage volume diffusion and pore elimination (Czerwinski, 2010).

7.1. Joints between magnesium alloys

It is found that diffusion bonding is one of the applicable procedures for joining of pure magnesium and its alloys. The experiments were performed at the pressure range of 2-20 MPa, temperature range from 300°C to 400°C and time periods up to 72 h. Combination of diffusion bonding with superplastic forming is often employed for manufacture of complex sheet structures. This combination allows to reduce weight and fabrication cost in comparison with mechanically fastened structures (Gilmore et al., 1991). The technology is successfully implemented for commercial wrought AZ31 magnesium alloy. The results of different experiments are summarized in Table 2.

Grain size (μm)	Bonding time (hour)	Bonding pressure (MPa)	Lap shear strength ratio
11	3	5	0.90
28	2	3	0.92
17	3	3	0.85
8.5	1	20	0.68

Table 2. Combination of diffusion bonding with superplastic forming which is used for superplastic wrought AZ31.

In general, the superplastic high temperature deformation mechanism and diffusion bonding have grain size dependency, as structures with finer grains will have lower bonding temperature (Somekawa et al., 2003).

8. Hybrid welding

Hybrid laser beam technologies are defined as a combination of a laser beam source with an additional secondary beam source or another joining technique. The use of hybrid heat sources for welding is relatively recent and the topic is of great interest. A hybrid laser-TIG welding (LATIG) of AZ31 alloy, gives higher welding speed than in laser or TIG welding (Liu et al., 2004) which a schematic of LATIC is shown in Fig. 24. The results found that joint penetration could be increased; arc stability improved and cost decreased.

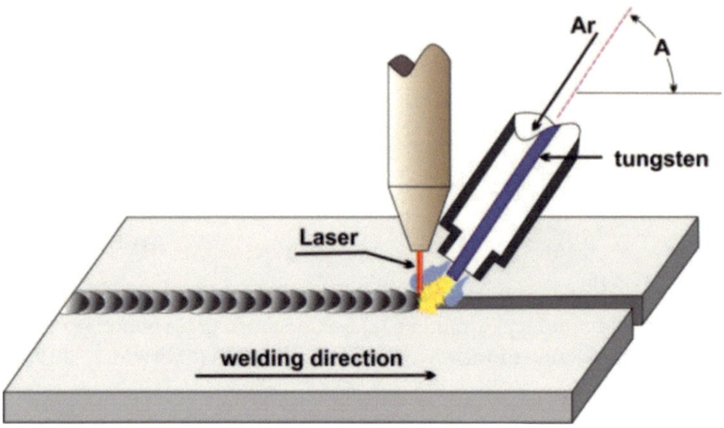

Figure 24. Principle of hybrid laser–TIG welding process.

The penetration depth is twice of that for TIG and four times of that for laser welding. A new method of hybrid joining, called laser continuous weld bonding, was developed as an alternative to laser welding and adhesive bonding (Liu and Xie, 2007). The technique was successful in joining of AZ31B magnesium and A6061 aluminium alloys by reducing a volume of brittle intermetallic compounds of Al_3Mg_2 and $Mg_{17}Al_{12}$ phases which are formed in the fusion zone, reducing the joint strength. The intermetallic phase formation was reduced due to the fluid generated by the gasification of adhesives. It appears that rising of adhesive vapour slows down the downward movement of liquid Mg, thus reducing its content in the weld. Hence, the weld is composed of two-phase mixture with less intermetallic compound and more solid solution.

9. Friction stir welding

In fusion welds the chemical composition of melt catches more attention, and fillers or consumable electrodes maybe are used. Also, the process maybe requires a neutral atmosphere to avoid the oxidation of melt. Sometimes surface preparation before welding is also essential. Several defects and difficulties due to the melting and solidification come to pass, resulting in weakness of mechanical properties such as tensile strength, fatigue properties and formability. From the other difficulties of these jointing methods, it can be implied to the porosity, oxidation, hot and cold cracks, and also inability of joint dissimilar metals with different thermal expansion coefficients.

"Solid state welds" are the processes in which the joint is formed below the melting temperature of material. Therefore, the oxidation does not occur and there is no need for shield gas, neutral atmosphere, and consuming material. Solid state welds consist of the processes such as friction, ultrasonic, forging, and explosion welding. Three major factors of Time, Temperature, and Pressure, solely or in combination can generate the weld joint.

Friction welding creates a joint without using filler metal, by converting the mechanical energy to the heat and the plastic deformation. Friction welding is divided into three categories: (i) rotational friction welding, (ii) non-rotational friction welding, and (iii) friction stir welding.

The rotational friction welding was the first friction welding which found commercially use. In this process, one of the cylindrical workpieces rotates concentrically to the other one, while two faces of cylinders are in contact. The material in the interface becomes soften, and then two cylinders are forged into each other (Fig. 25). This welding method is used for similar and dissimilar materials.

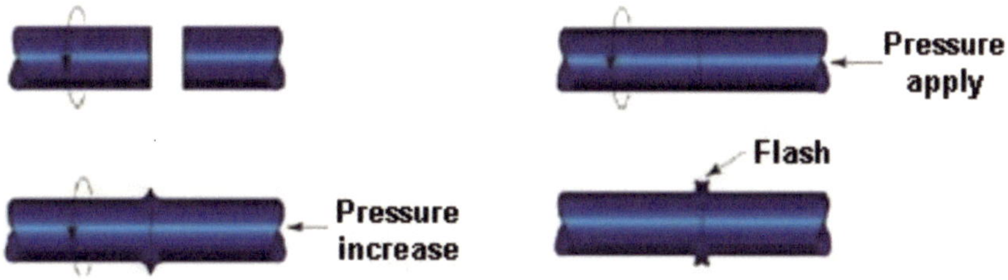

Figure 25. Schematic illustration of rotational friction welding.

In non-rotational friction welding the linear, orbital or angular movements are used to join the rectangular straps.

Friction stir welding (FSW), as a solid-state joining process, is invented by Wayne Thomas and his colleagues at The Welding Institute of UK in 1991. Although it was primarily applied on aluminium, it is now widely used for joining dissimilar metals together such as different aluminium and also magnesium, copper and ferrous alloys. FSW is believed to be the most important development in joining of metals in the past two decades.

The difficulty of making high-strength, fatigue and fracture resistant welds in aerospace alloys especially aluminium alloys, such as highly alloyed 2XXX and 7XXX series, has restricted the wide use of welding for joining aerospace structures. These aluminium alloys are generally classified as non-weldable, because of the poor solidification microstructure and porosity in the fusion zone. Also, the weakness of mechanical properties compared to the base metal is very noticeable. These factors have made the welding of these alloys unpleasant by the conventional methods.

Thus, in friction stir welding, which is a developed form of friction welding, the major aim is to join the materials which are hard to joint by the fusion welding processes. In the conventional friction welding, the heat is generated by friction between to faces in the joining interface. However, in FSW a third substance as a non-consumable, wear-resistant, rotating tool is inserted into the mated edges of jointing sheets and traversed along the joint line (Fig. 26a). The tool includes two main parts: (i) shoulder, to produce the frictional heat and support the softened material; and (ii) pin, to stir the material. Fig. 26b and 26c show a simple and a specially designed FSW tool. The pin and the shoulder possess special designs, which govern the process parameters and also resultant weld properties. Exactly, the tool provides two primary functions: (i) heating of workpiece, which is induced by friction between the tool and the workpiece, and also plastic deformation of the material; and (ii) movement of material to create the weld. The material around the pin is softened by concentrated heat, and moved from the front to the back of the pin by combination of tool rotation and translation. Finally, the moved material to the back of the pin is forged to weld nugget by the tool shoulder, resulting in a fine joint. Due to the various geometrical features of the tool, the material flow around the pin is quite complex. During FSW, the material undergoes an intense plastic deformation at the high temperature, resulting in a recrystallized microstructure with the fine equiaxed grains, and consequently good mechanical properties (Mishra and Ma, 2005).

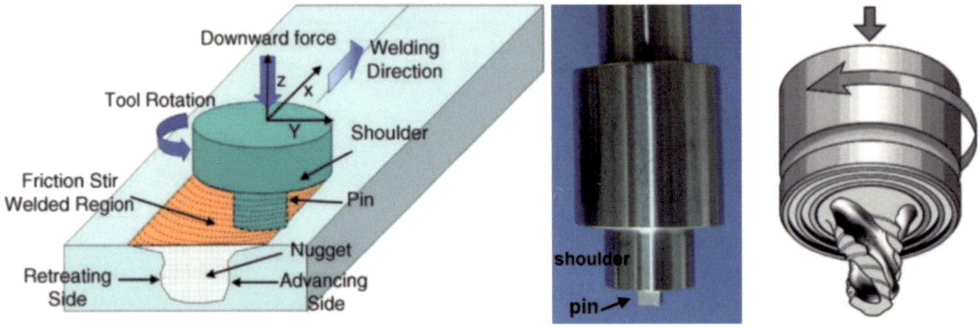

(a) (b) (c)

Figure 26. (a) Schematic illustration of FSW, (b) a simple FSW tool, and (c) an especially designed FSW tool.

Because of energy efficiency, environmental friendliness, and versatility, FSW is called as a green technology. In comparison with the conventional welding processes, FSW consumes significantly less energy, i.e., only 2.5% of the energy required for laser welding. No shielding gas or flux is needed and the poisonous gases are not generated, thus the process is nature friendly. The tensile strength and fatigue properties of products are appropriate and no filler metal is needed to use. Dissimilar aluminium and magnesium alloys, and composites can be joined with equal ease, and the thermal conduction coefficient is less important in compare with the fusion welding processes (Mishra and Ma, 2005). Contrary to the conventional friction welding, which is usually applied on small axisymmetric parts, FSW can be implemented for various types of joints like butt, lap, T butt, and fillet joints. As a result of the low heat experienced by the material, the products have excellent dimensional stability and repeatability, and also there is no loss of the alloying elements.

Since there is no melting of the material in FSW, fewer defects than the fusion welding processes are seen. In this process the HAZ, presenting many defects and the main reason for deterioration of mechanical properties, is very narrow. Due to the fact that the FSW does not involve any melting pool, the FSW is not limited to an especial position, and jointing sheets can be welded in any position. The process is simple and, in contrast to the fusion welds, does not need a proficient operator. This welding method is not sensitive to surface quality, and thus there is no need to degrease or remove the oxide layers.

9.1. Process parameters

FSW includes complex material flow which is induced severely by welding parameters and tool geometry. These parameters significantly affect temperature distribution, and therefore the microstructural characteristics of material.

9.1.1. Tool geometry

Tool geometry is the most consequential factor on the weld quality and has a crucial role in material flow. As declared before, the tool performs two primary functions of localized heating, and material flow. The tool is plunged till the shoulder touches the workpiece. The friction between the tool (shoulder and pin) and workpiece produces the biggest portion of heating. It is clear that, the amount of produced heat will increase as the shoulder diameter increases. The shoulder also supports the softened material during the process. The second task of the tool is to 'stir' and 'move' the softened material. This role commences when the tool travel starts along the welding line. In this case material undergoes a severe plastic deformation which results in an additional heating. The uniformity of microstructural and mechanical properties as well as process loads is governed by the tool design. Generally a concave shoulder and threaded cylindrical or square pins are used.

9.1.2. Welding parameters

Two process parameters in FSW are very important: tool rotational speed, and tool traverse speed along the welding line. The tool rotation results in stirring and mixing of softened materials around the pin, and the tool translation moves them from the front to the back of the pin and finishes the weld. Higher tool rotational speed to traverse speed ratio (ω/v) induces higher temperature due to the higher frictional heat, and results in stronger stirring of material. On the other hand, grain size in the SZ becomes larger with increasing ω/v, due to the increase in heat input, promoting the growth of recrystallized grains.

Besides the tool rotational and traverse speeds, another important process parameter is the "tilt angle" or "tool tilt" with regard to the workpiece surface. An appropriate tilt angle towards traverse direction ensures that the tool shoulder supports the stirred materials and moves them properly from the front to the back of the pin, and forges them into the weld nugget. Tilt angle can be adjusted between 0 and 6°.

Further, the tool penetration depth into the workpiece is another important factor to produce sound welds. When the penetration depth is too shallow, the contact surface between the tool shoulder and the workpiece is insufficient. Thus, the material will not be softened adequately, and then the tool will not be able to move the material efficiently, resulting in creation of tunnelling cavity or surface groove. On the other hand, if the penetration depth is too deep, the tool shoulder plunges into the workpiece, generating immoderate flash. With respect to the tool tilt angle, a sound weld can be achieved when one third to two third of the tool shoulder touches the workpiece surface.

9.2. Microstructural characteristics

In the cross section of friction stir welded parts, three distinct zones based on microstructural characteristics of grains and precipitates can be identified. As shown in Fig. 27 these zones are consisting of: stirred zone (SZ) (is called also nugget zone), thermo-mechanically affected zone (TMAZ), and heat affected zone (HAZ). Severe plastic deformation accompanied by high temperature during FSW results in recrystallization, grain refinement, and precipitate dissolution within the SZ (Mishra and Ma, 2005).

FSW causes the temperature to increase up to 0.6-0.9 of melting temperature (MT) of work-piece within the SZ. At such a high temperature, precipitates in alloys can coarsen or dissolve into the matrix depending on alloy type and peak temperature. Generally it can be mentioned that, a combination of dissolution, coarsening and/or reprecipitation of strengthening precipitates during FSW is possible to occur.

Unique to the FSW process, the TMAZ, which is created between the parent material and the SZ, experiences both high temperature and plastic deformation during the process. A typical TMAZ, characterized by a highly deformed structure, is shown in Fig. 28. This figure is related to the AZ91 magnesium alloy. As seen in the figure, α and β phases have found an special orientation, and are prolonged within the boundary of the SZ and the BM. Recrystallization does not occur entirely in the TMAZ due to insufficient deformation strain and lower temperature, however dissolution of some precipitates or maybe partial recrystallization can be observed in the TMAZ.

As shown in Fig. 28, the basic microstructure of AZ91 magnesium alloy consists of primary α-phase in which aluminium rich β-phase ($Mg_{17}Al_{12}$) is precipitated along the grain boundaries. While the average grain size is about 150 μm in the base metal, it is about 10 μm in the SZ.

Figure 27. Various microstructural zones in the cross section of friction stir welded AZ91 alloy.

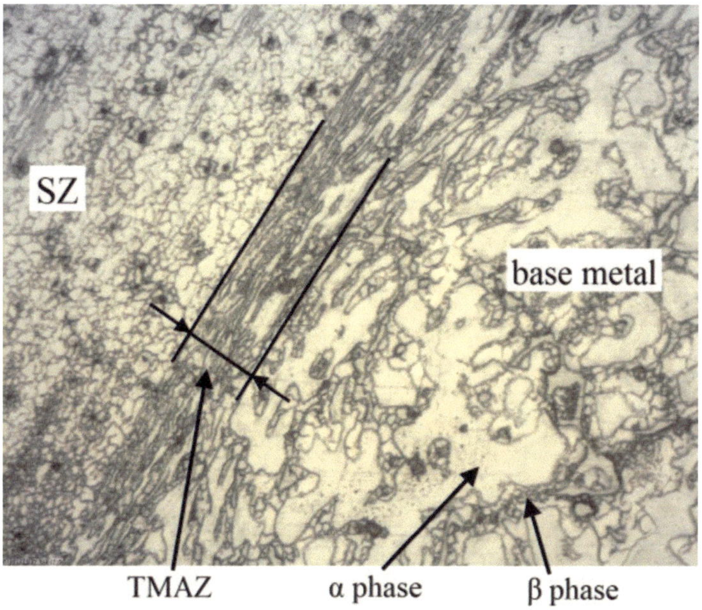

Figure 28. SZ, TMAZ, and base metal in friction stir welded AZ91 magnesium alloy.

Heat-affected zone, beyond the TMAZ, experiences a thermal cycle, but not any plastic deformation. The grain structure of the HAZ is as same as the parent material; however the thermal exposure above 0.4 TM can exert an effect on the precipitate structure depending on alloy type.

9.3. FSW of Magnesium Alloys

Magnesium alloys, consisting hexagonal-closed pack structure, have poor formability, and sheet material of them is made by casting processes, except some wrought alloys such as AZ31. Therefore, it is almost difficult to weld these cast alloys by conventional methods due to the porosity formation. Furthermore, relatively high expansion coefficient of magnesium causes large deformation/distortion of product. Hence, FSW as a solid-state welding technique can be the optimum choice for welding cast magnesium alloy sheets. FSW studies, performed on different magnesium alloys, come together following considerable observations.

The quality of FSW of magnesium alloys is significantly sensitive to the tool rotational and traverse speeds which determine the amount of produced heat. It is reported that optimum parameter for FSW of AZ91 sheets are limited to a narrow range, i.e., higher tool rotational speeds and lower traverse speeds. Butt welding using square-pin tool can successfully perform at the optimum parameter combinations of traverse speed of 50 mm/min and tool rotational speed of 900–1800 rpm. By decreasing the rotational speed, the traverse speed should be decreased relatively. Furthermore, increasing the shoulder to pin diameter rate can expand the optimum parameters range. Higher traverse speeds or lower rotational speeds than the optimum parameters caused the formation of tunnelling cavity or a lack of material bonding in the SZ. This is because of the poor formability of cast AZ91 magnesium alloy with a lot of intermetallic compounds, β-$Mg_{17}Al_{12}$, at the grain boundaries. Additionally, in order to evade sticking of magnesium to the tool during FSW, a relatively high rotational speed is utilized (Suhuddin et al., 2009). Similarly, for hot-rolled AZ31B-H24 reported that sound joints were produced only at higher ω/v rates.

FSW of magnesium alloy usually does not generate a liquid phase (Mishra et al., 2005). Based on microstructural characteristics, the peak temperature is predicted to be between 370 and 500 °C within the SZ. On the other hand, simulation results show that the peak temperature in the SZ is about 380 to 550 °C depending on process parameters and tool design (Asadi et al. 2011a). However, a relative low tool traverse and high rotational speed can cause generation of liquid phase and a complex microstructure, especially in a thin layer at the top of the weld zone.

According to the Mg–Al binary phase diagram, β-$Mg_{17}Al_{12}$ intermetallic compounds in AZ91 will dissolve into the magnesium matrix completely when the heating temperature is higher than 370 °C. This critical temperature is 320 °C for the AZ61 and AM60 alloys. For as-cast Mg-Al alloys, the coarse continuous β-$Mg_{17}Al_{12}$ intermetallic phases disappear after FSW (see Fig. 28) and their particles exist both inside the grains and at the grain boundaries. These particles are spherical- or ellipsoidal-shaped, with the size range of 20 to 100 nm (Zhang et al., 2005). Although the peak temperature of SZ is high enough for $Mg_{17}Al_{12}$ to be dissolved into the magnesium matrix, but the temperature is not the entire problem (Zhang et al., 2005). Studies about the influence of heat treatment on microstructure revealed that, although partial dissolution of intermetallic compounds occur, most of the network structure (about 70%) was remained untouched, after being heated at 350 °C for 1 hour (Mishra et al., 2005). Indeed, dissolution and re-precipitation could not be the main reason for the morphology change of intermetallic compound, since the thermal history of FSW is not adequate for this mechanism. On the other hand, the size and shape of intermetallic compounds changed mainly through mechanical fracture. The broken particles are distributed in the matrix of SZ with the material flow caused by tool rotation. Partial dissolution and coarsening will help the broken particles to further spheroidize in some degree (Zhang et al., 2005).

The hardness of the stirred zone depending on, process parameters, grain size, intermetallic components, and the way that the base metal is produced, can be higher or lower than the base metal. Variation of hardness with grain size is proved to follow the Hall–Petch relationship, as hardness increases with decreasing grain size. However, a reduction in hardness is revealed in the SZ, due to coarsening of grains and/or dissolution of strengthening precipitates such as $Mg_{17}Al_{12}$.

FSW improves tensile properties of cast magnesium alloys such as AZ91 (Asadi et al., 2011b and 2012), while a reduction in tensile properties is observed in wrought magnesium alloys such as AZ31B-H24 and AZ61 (Mishra et al., 2005). For the FSW of AZ91D it is reported that in the case of transverse tensile test (perpendicular to weld line), all of the test specimens fractured in the base metal (Mishra et al., 2005), indicating the joint efficiency of 100% for these FSW joints. Longitudinal tensile tests revealed that the strength and elongation of the SZ were considerably enhanced compared to those of the base metal. In contrary, a reduction in tensile properties of FSW of AZ31B-H24 with a fracture close to the SZ in the transition region (TMAZ) is reported. Generally, it can be concluded that the joint efficiency of FSW for magnesium is very high (80–100%). This means that, FSW is potentially excellent process for the joining of magnesium alloys (Suhuddin et al., 2009).

Additionally, by inserting some ceramic powders in the interface of jointing sheets, it is possible to produce a metal-matrix-composite weld. In this case, strength of weld nugget increases, but its elongation decreases compared to the case without ceramic powder.

One of the most important advantageous of FSW is possibility of making dissimilar joints. Typically, in dissimilar friction stir welding between 5052 aluminium alloy and AZ31 magnesium alloy, it is reported that the peak hardness in the SZ is twice higher than that of the base metals. The Vickers microhardness test was carried out along the dashed lines marked in Fig. 29a, which were 1.5 mm (top), 3 mm (middle) and 4.5 mm (bottom) to the upper surface, and the results are presented in Fig. 29b. The sharp fluctuations in hardness of the SZ are attributed to the onion ring structure and the intercalated microstructure. The tensile fracture position locates at the advancing side (aluminium side), where the hardness distribution of weld shows a sharp decrease from the stir zone to 5052 base material (Yong et al., 2010).

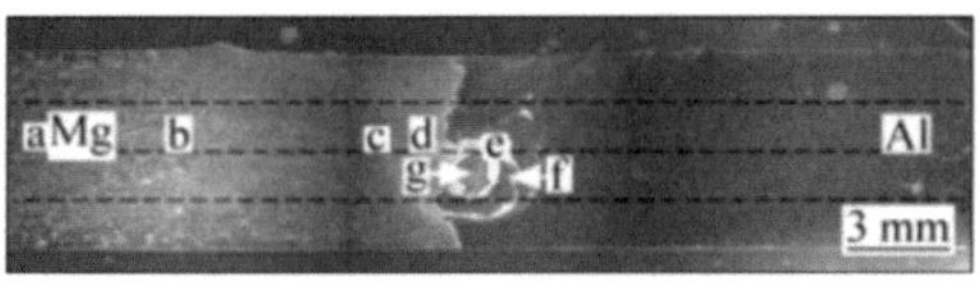

(a)

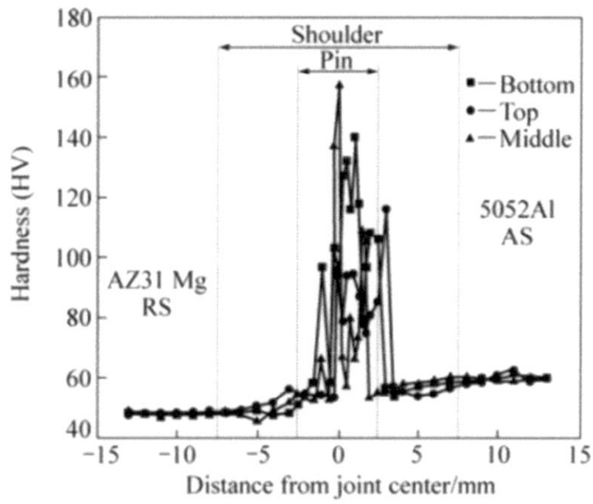

(b)

Figure 29. (a) Optical macrograph, and (b) Microhardness profiles of cross-section of dissimilar weld (Yong et al., 2010).

10. Summary

During last decade, a substantial proceed is made in welding and joining of magnesium alloys. To improve in conventional fusion welding techniques, new methods and their hybrids are developed. On the other hand, using non-fusion welding methods can also asset decreasing some metallurgical defects. Among the different welding processes for magnesium and its alloys, laser welding and friction stir welding will probably become efficient joining techniques. Crack-free laser welded joints with low porosity and good surface quality can be obtained for some magnesium alloys, in particular wrought material. Friction stir welding has overcome to most of the problems arose in the fusion welds. However, its drawbacks such as limited weld designs, high mechanical forces to jointing sheets, and thus requiring stronger fixturing may have confined its use. Indeed, joining techniques, which require local deformation, are not fully successful methods due to limited formability of magnesium, especially casting alloys such as AZ91. In this case, scientific investigations are still needed to understand and overcome these basic weldability problems of magnesium alloys.

Author details

Parviz Asadi
School of Mechanical Engineering, College of Engineering, University of Tehran, Tehran, Iran

Kamel Kazemi-Choobi
Research Center for Advanced Materials and Mineral Processing, Faculty of Materials Engineering, Sahand University of Technology, Tabriz, Iran

Amin Elhami
Department of Mechanical Engineering, Tehran Central Branch, Islamic Azad University, Tehran, Iran

11. References

Aizawa T. (2003). Magnetic pressure seam welding method for aluminium sheets, *Welding Int.*, Vol. 17, pp. (929–933).

Asadi P., Mahdavinejad R.A. and Tutunchilar S. (2011). Simulation and experimental investigation of FSP of AZ91 magnesium alloy, *Materials Science and Engineering A*, Vol. 528, pp. (6469– 6477).

Asadi P., Faraji G., Masoumi A. and Besharati Givi M. K. (2011). Experimental Investigation of Magnesium-Base Nanocomposite Produced by Friction Stir Processing: Effects of Particle Types and Number of Friction Stir Processing Passes, *Metallurgical and Materials Transaction A*, Vol. 42A, pp. (2820-2832).

Asadi P., Besharati Givi M. K., Parvin N., Araei A. Taherishargh M. and S. Tutunchilar, (2012). On the role of cooling and tool rotational direction on microstructure and mechanical properties of friction stir processed AZ91, *International Journal of Advanced Manufacturing Technology*, DOI 10.1007/s00170-012-3971-0.

ASM (1990) *Metals Handbook*, ASM International, Materials Park, Ohio

ASM (1993) *Welding Brazing and Soldering*, ASM International, Materials Park, Ohio

Berlin A. (2011). *Magnetic Pulse Welding of Mg Sheet,* MSC Thesis, Waterloo, Ontario, Canada.

Berlin A., Nguyen T.C., Worswick M.J. and Zhou Y. (2011). Metallurgical analysis of magnetic pulse welds of AZ31 magnesium alloy, *Science and Technology of Welding and Joining*, Vol. 16, pp. (728).

Busk, R. S. (1987). *Magnesium Products Design*, Marcel Dekker Inc., ISBN, NewYork.

Cao X., Jahazi M., Immarigeon J. P. and Wallace W. (2006). A review of laser welding techniques for magnesium alloys, *Journal of Materials Processing Technology*, Vol. 171, pp. (188–204).

Cao X., Xiao M., Jahazi M. and Immarigeon J. -P. (2005). Continuous Wave ND:YAG Laser Welding of Sand-Cast ZE41A-T5 Magnesium Alloys, *Materials and Manufacturing Processes*, Vol. 20, pp. (987–1004).

Chi C. T. and Chao C. G. (2007). Characterization on electron beam welds and parameters for AZ31B-F extrusive plates, *Journal of Materials Processing Technology*, Vol. 182, pp. (369–373).

Chi C. T., Chao C. G., Liu T. F. and Wang C. C. (2006). A study of weldability and fracture modes in electron beam weldments of AZ series magnesium alloys, *Materials Science and Engineering A*, Vol. 435–436, pp. (672–680)

Czerwinski F. (2010). *Welding and joining of magnesium alloys*, In: Magnesium Alloys - Design, Processing and Properties, chapter 21, pp. (469-490).

Dhahri M., Masse J. E., Mathieu J. F., Barreau G. and Autric M. (2000). CO_2 laser welding of magnesium alloys, X. Chen, T. Fujioka, A. Matsunawa (Eds.), *Proceedings of the SPIE: High-Power Lasers in Manufacturing*, Vol. 3888, 2000, pp. (725–732).

Dhahri M., Masse J. E., Mathieu J. F., Barreau G. and Autric M. (2001a) Laser weldability of WE43 magnesium alloy for aeronautic industry, *Proceedings of the Third LANE2001: Laser Assisted Net Shape Engineering 3*, Erlangen, 28–31 August 2001.

Dhahri M., Masse J. E., Mathieu J. F., Barreau G. and Autric M. (2001b). Laser welding of AZ91 and WE43 magnesium alloys for automotive and aerospace industries, *Advance Engineering Materials*, Vol. 3, No. 7, pp. (504–507).

Dilthey U., Haferkamp H., Niemeyer M. and Trager G. (1998). Laser and EB welding of magnesium alloys, *American Institute of Welding*, IIW Document No. IV-701-98.

Gilmore C. J., Dunfold D. V. and Parteidge P. G. (1991). *Journal of Materials Science*, Vol. 26, pp. (3119-3124).

Haferkamp H., Bach Fr. -W., Burmester I., Kreutzburg K. and Niemeyer M. (1996). Nd:YAG laser beam welding of magnesium constructions, *Proceedings of the Third International Magnesium Conference, UMIST*, Manchester, UK, 10–12 April 1996.

Haferkamp H., Goede M., Bormann A. and Cordini P. (2001). Laser beam welding of magnesium alloys–new possibilities using filler wire and arc welding, *Net Shape Engineering*, Vol. 3, pp. (333–338).

Haferkamp H., Niemeyer M., Dilthey U. and Trager G. (2000). Laser and electron beam welding of magnesium materials, *Weld. Cutt.*, Vol. 52, No. 8, PP. (178–180).

Haferkamp H., Niemeyer M., Schmid C., Kaese V. and Cordini P. (2000). Laser welding of magnesium alloys–cooling conditions and resulting metallurgical properties magnesium 2000, *Second International Conference On Magnesium Science and Technology*, Dead Sea, Israel, 22–24 February 2000.

Hwang I. S., Kim D. C. and Kang M. J. (2011). Inverter DC resistance spot welding of magnesium alloy AZ31, *Materials Manufacturing and Processing*, Vol. 48, No. 2, pp. (112-117).

Hwang I. S., Yoon H. J., Kang M. J. and Kim D. C. (2010). Weldability of 440 MPa galvanized steel with inverter DC resistance spot welding process, *Journal of Achievements in Materials and Manufacturing Engineering*, Vol. 42, No. 1, pp. (37-44).

Kulekci, M. K. (2008). Magnesium and its alloys applications in automotive industry, *International Journal of Advanced Manufacturing Technology*, Vol. 39, pp. (851–865), ISSN.

Kore S. D., Imbert J., Worswick M. J. and Zhou Y. (2009). Electromagnetic impact welding of Mg to Al sheets, *Science and Technology of Welding and Joining*, Vol. 14, pp. (549).

Kou S. (2003). *Welding Metallurgy* (second edition), John Wiley & Sons, Inc, 0-471-43491-4, Hoboken, New Jersey.

Lathabai S., Barton K. J. and Harris D. (2003). Welding and weldability of AZ31B by gas tungsten arc and laser beam welding processes, *Magnesium Technology*, Vol. 157.

Lehner C., Reinhart G. and Schaller L. (1999). Welding of die cast magnesium alloys for production, *Journal of Laser Applications*, Vol. 11, No. 5, pp. (206–210).

Leong K. H. (1998). Laser beam welding of AZ31B-H24 magnesium alloy, *Proceedings of Laser Institute of America, ICALEO' 98 Conference*, Orlando, FL, USA, 1998.

Liu L. and Xie L. (2007). Adhesive bonding between Mg alloys and polypropylene, Materials Technology: *Advanced Performance Materials*, Vol. 22, No. 2, pp. (76-80).

Liu L., Wang J. and Song G. (2004). Hybrid laser-TIG welding, laser beam welding and gas tungsten arc welding of AZ31B magnesium alloy, *Materials Science and Engineering A*, Vol. 381, pp. (129-133).

Luo Y., You G., Ye H. and Liu J. (2010). Simulation on welding thermal effect of AZ61 magnesium alloy based on three-dimensional modeling of vacuum electron beam welding heat source, *Vacuum*, Vol. 84, pp. (890-895).

Marya M. and Edwards G. R. (2000). The laser welding of magnesium alloy AZ91, *Welding World*, Vol. 44, No. 2, pp. (31–37).

Marya M. and Edwards G. R. (2001). Factors Controlling the Magnesium Weld Morphology in Deep Penetration Welding by a CO2 Laser, *Journal of Materials Engineering and Performance*, Vol. 10, No. 4, pp. (35).

Marya M. and Edwards G. R. (2002). Chloride contribution in flux-assisted GTA welding of magnesium alloys, *Welding Journal*, Vol. 12, pp. (291-298).

Marya M. and Edwards G. R. (2002). Influence of laser beam variable on AZ91D weld fusion zone microstructure, *Science and Technology of Welding and Joining*, Vol. 7, No. 5, pp. (286–293).

Marya M., Olson D. L. and Edwards G. R. (2000). Welding of magnesium alloys for transportation applications, *Proceedings from Materials Solution '00 on Joining of Advanced and Specialty Materials*, ISBN, St. Louis, Missouri, 9–11 October 2000.

Mishra R. S. and Ma Z. Y. (2005). Friction stir welding and processing, *Materials Science and Engineering R*, Vol. 50, pp. (1–78).

Mordike B. L. and Ebert T. (2001). Magnesium: properties-applications-potential, *Materials Science and Engineering A*, Vol. 302, pp. (37–45).

Munitz A., Cotler C., Stern A. and Kohn G. (2001). Mechanical properties and microstructure of gas tungsten arc welded magnesium AZ91D plates, *Materials Science and Engineering A*, Vol. 302, pp. (68–73).

Oates W. R. (1996). *Welding Hand book* (eighth edition), American Welding Society, Miami, Florida.

Pastor M., Zhao H. and DebRoy T. (2000). Continuous wave Nd:yttrium–aluminium–garnet laser welding of AM60B magnesium alloys, *Journal of Laser Applications*, Vol. 12, No. 3, pp. (91–100).

Powers D. E. and Laflamme G. R. (1992). *Welding Journal*, pp. (47–52).

Shi H., Qiu R., Zhu J., Zhang K., Yu H. and Ding G. (2010). Effects of welding parameters on the characteristics of magnesium alloy joint welded by resistance spot welding with cover plates, *Materials and Design*, Vol. 31, pp. (4853-4857).

Somekawa H., Watanabe H. and Higashi K. (2003). The Grain Size Dependence on Diffusion Bonding Behavior in Superplastic Mg Alloys, *Materials Transactions*, Vol. 44, No. 4 pp. (496-503).

Stolbov V. I., El'tsov V. V., Oleinik I. A. and Matyagin V. F. (1990). Effect of the nature of thermal processes on cracking in repair welding components of magnesium alloys, *Svarochnoe Proizvodstvo*, Vol. 37, No. 5, pp. (29–31).

Suhuddin U. F. H. R., Mironov S., Sato Y. S., Kokawa H. and Lee C. -W. (2009). Grain structure evolution during friction-stir welding of AZ31 magnesium alloy, *Acta Materialia*, Vol. 57, pp. (5406–5418).

Wang Z., Gao M., Tang H. and Zeng X. (2011). Characterization of AZ31B wrought magnesium alloy joints welded by high power fiber laser, *Materials Characterization*, Vol. 62, pp. (943 – 951).

Xiaoa L., Liua L., Chenb D. L., Esmaeili S. and Zhou Y. (2011). Resistance spot weld fatigue behavior and dislocation substructures in two different heats of AZ31 magnesium alloy, *Materials Science and Engineering A*, Vol. 529, pp. (81–87).

Yin Y. H., Sun N., North T. H. and Hu S. S. (2010). Microstructures and mechanical properties in dissimilar AZ91/AZ31 spot welds, *Materials Characterization*, Vol. 61, No. 10, pp. (1018-1028).

Yong Y., Da-tong Z., Cheng Q. and Wen Z. (2010). Dissimilar friction stir welding between 5052 aluminum alloy and AZ31 magnesium alloy, *Transition of Nonferrous Metal Society*, China, Vol. 20. pp. (619-623).

Yu L., Nakata K., Yamamoto N. and Liao J. (2009). Texture and its effect on mechanical properties in fiber laser weld of a fine-grained Mg alloy, *Materilas Letters*, Vol. 63, pp. (870–872).

Zhang D., Suzuki M. and Maruyama k. (2005). Microstructural evolution of a heat-resistant magnesium alloy due to friction stir welding, *Scripta Materialia*, Vol. 52, pp.)899–903).

Zhang Z. D., Liu L. M., Shen Y. and Wang L. (2008). Mechanical properties and microstructures of a magnesium alloy gas tungsten arc welded with a cadmium chloride flux, *Materials Characterization*, Vol. 59, pp. (40 – 46).

Zhao H. and Debroy T. (2001). Pore formation during laser beam welding of die cast magnesium alloy AM60B–mechanism and remedy, *Welding Journal*, Vol. 80, No. 8, pp. (204–210).

Zhu J., Li L. and Liu Z. (2005). CO2 and diode laser welding of AZ31 magnesium alloy, *Applied Surface Science*, Vol. 247, pp. (300–306).

Section 6

Developments on Magnesium Alloys Applied to Transport

Chapter 7

Application of Magnesium Alloys in Transport

W.A. Monteiro, S.J. Buso and L.V. da Silva

Additional information is available at the end of the chapter

http://dx.doi.org/10.5772/48273

1. Introduction

Magnesium alloys have always been attractive to designers due to their low density, only two thirds that of aluminum alloys in the aerospace industry and therefore can be an innovation technology if used for low weight airframe structures. This has been a major factor in the widespread use of magnesium alloy castings and wrought products.

New light materials are currently inserted in world strategies of transport vehicle industry since the environment necessities for pollution and reduction of fuel consumption. Therefore the industry takes part of the risk of development of such alloys but, in fact, some of this has been made at academic level. Some aspects of the necessities and characteristics concerning those alloys are: low costs, insulation (sound and thermal), impact safety, deformation strength, recyclability and guarantee (to aging as example). All those aspects are linked with the increasing of new vehicle models and reflect in production programs that are more and more complexes (Raynor, 1959; Roberts, 1960; Eliezer et al, 1998).

However to use this low weight material several mechanical properties have to be increased and the technological behavior improved. A further requirement in recent years has been for superior corrosion performance and dramatic improvements have been demonstrated for new magnesium alloys. Improvements in mechanical properties and corrosion resistance have led to greater interest in magnesium alloys for aerospace and specialty applications.

Magnesium alloys are also used in many other engineering applications where having light weight is a significant advantage. Magnesium-zirconium-aluminum alloys tend to be used in relatively low volume applications where they are processed by sand or investment casting, or wrought products by extrusion or forging.

Actually it´s well known that there are a growth in primary production of magnesium due to a supplementary application of your products in market. A total 2010 production of primary magnesium in the world was approximately 809,000 tons. The main producer are

China 654,000 tons, USA 45,000 tons, Russia 40,000 tons, Israel 30,000 tons, Kazakhstan 20,000 tons and Brazil 16,000 tons (China Magnesium Association; U. S. Geological Survey).

2. Magnesium alloys for transport applications

The automobile have more than a hundred years since its invention and the light alloys have been utilized since 1915. In the 30's the aluminum did substitute almost completely the casting iron as main component of pistons. In the same decade, in Germany, magnesium alloys were utilized in the production of camshaft and in the gear box, which leads, at that time, the total decrease of weight at least 7% of the total of the automobile.

Magnesium alloy developments have traditionally been driven by aerospace industry requirements for lightweight materials to operate under increasingly demanding conditions. In the last decade several heavy magnesium parts have been assembled in passenger cars, such as gear box housings and crank cases (Aghion, 2003).

Considerable research is still needed on magnesium processing, alloy development, joining, surface treatment, corrosion resistance, and mechanical properties improvement. Surface coatings produced for magnesium die-casting by hexavalent chromium baths have been used to provide stand-alone protection and as a pretreatment for painting. Teflon resin coating has been developed for Mg alloys; initially the coating is obtained with an aluminum vapor deposition and finish treatment with a Teflon resin coating. The Teflon resin coating is a low cost, chromium-free corrosion resistant coating for magnesium alloys. The coating not only has corrosion resistant properties, but also good lubricity, high frictional-resistance and non-wetting properties (Kulekci, 2008).

These new projects are concerned on the modification of existing and the development of new magnesium wrought products (sheets and extrusions), that provide significantly improved static and fatigue strength properties for lightweight fuselage applications (Development of material models and failure criteria for the prediction of forming processes, plastic deformation and failure behavior of components; material adapted design and the evaluation of structural behavior to close the process and development chain for aeronautic components).

The specific strength properties of these innovative materials are required to be higher than AA2024 for structural applications (secondary structure) and higher than AA5083 for non-structural applications. At the beginning of the project new alloys will be developed and experimental alloys will be tested. Appropriate manufacturing (rolling, extrusion), forming and joining technologies require development, simulation and validation for the innovative material and application (Hombergsmeier, 2011). The technological objective is a weight reduction of the fuselage structure, system and interior components up to 30 %. The strategic objectives are an increase in the operational capacity of 10 %, a reduction in the direct operating cost of 10 % and finally a reduction in the fuel consumption of 10 % and therefore a reduced environmental impact with regard to emissions and noise (Hombergsmeier, 2009).

Regarding corrosion is also a problem to be solved with newly adapted surface protection systems according to aerospace requirements and advanced design concepts. The addition of chemical elements and special surface treatments in magnesium to avoid flammability (auto-ignition temperature); Mg alloys with eutectic phases, low melting temperature, ignite first with comparative investigations to aluminum alloys (Hombergsmeier, 2011).

For those purposes there are nowadays some new techniques for the production of this kind of light alloys, notably: powder metallurgy (PM), which is essentially the process that leads small particles (from sizes in a range from nanometers to tens of micrometers) to a consolidation process that results in a uniform and fine grained structure material; spray forming techniques that relies in production processes which allows the production of parts and components by deposition of sprayed particles in semi-molten state that will consolidate in an injection mold with a substrate and metal foam. In addition the nanostructured materials have been the utmost in materials development due mainly to their particular relationship between microstructure of grains or hyperfine particles and mechanical properties, differentiated from those produced by conventional processes.

The development of solid state powder metallurgy makes it possible to increase the values of solute above the limits of conventional metallurgy arising a new era for the magnesium and aluminum alloys. Increasing the amount of "thirds" alloying elements gives to these alloys new characteristics such as high deformation rates at room temperature, refinement of microstructure (hyperfine grains or nanostructured materials), new conditions for cold and hot work processing, welding, wear resistance, increase of hardness as examples.

Also, recyclability of magnesium and aluminum alloys is a growing factor in choice for utilization in automotive industry, since nowadays it's possible to recycle up to 99% of parts and pieces made of magnesium and aluminum which reduces the costs with transportation and this is also good for the environment.

News production technologies search for minimize the quantities of energy employed in its processes and the recyclability shows an economy of 95% of energy comparing to that produce primary magnesium and aluminum generating an "ecological compensation".

Another influence in research for new technologies of materials and light alloys relies on the shortening of life of models. New materials and alloys are usually more expensive than commercial materials, so there is direct needs of investment in reduce the costs of production and development of such alloys allowing the utilization in large scale in the automobile production processes.

Despite the fact that the introduction of light alloys and new technology light alloys is a tendency not changeable, the utilization effective today still is more applicable to competition or sportive cars and motorcycles, due to the high costs previously mentioned. However if compared along the existence of automobile the employment of light alloys rise exponentially from the earliest up to the latest commercial model. The influence of 1970's in the development of such technologies is notable comparing with the few kilograms used in the first automobiles. All those factors contribute to new researches and development of this class of materials for structural and mechanical applications in automotive industry (transport).

Global effort in magnesium materials development has diversified in recent years, targeting new automotive, aerospace, and also in biomedical applications. The 1990s saw a strenuous effort in the tissue to regenerate without the use of a second surgery to remove the repair implant. Research in this field is focusing on the development of new magnesium alloys with low rates of bio-degradation. Emerging areas of application for magnesium are the input for global effort in magnesium materials development. Interdisciplinary and international collaborations will be the key in developing magnesium to its full potential as an environmentally friendly structural material with wide-ranging applications (Beals et al., 2007).

A very interesting program to benefit the magnesium utilization is MagForming, an international project (12 different partners) with the aim to advance the state of the art in forming methods for a range magnesium alloys in extruded, sheet and plate forms using aerospace prototypes. This includes forging, superplastic forming, roll bending, pad forming, deep drawing and creep forming. The project aims to develop best practices for all of these methods via production of aerospace prototype parts. The data results in production of useful parts for testing, potentially leading to the use of magnesium in the system and structure areas from which the parts have been taken (Davis et al., 2009).

As mention earlier, magnesium alloys had been widely used in aircrafts up to the 1950s. Due to some major drawbacks such as high corrosion sensitivity and, under extreme conditions, flammability other materials, both metals and plastics, have increasingly replaced magnesium. Airbus is now reappraising the introduction of magnesium alloys in aircraft based on the results of recent research projects that suggest some promising applications for different alloys and product forms. Additionally Airbus is developing and investigating approaches for a lab scale fire testing procedure with the scope to ensure compliance with FAA-regulations on aircraft fire safety (Knüwer at al. 2009).

3. The future of light alloys application including environment aspects

New energies that will substitute the mineral energy font (petroleum) and also nowadays the increasing emission of CO_2 in the entire world in the last decades lead to a special condition. So new regulations attempt to control or reduce emission and consumption has been created all over the world. In Europe the requirements also faced the reduction of production and maintenance costs and enhance the lifetime and reliability of automobile. All these conditions bring a wide field of possibilities for light alloys and new processes of production (Ditze, 1999; Kimura, 2002; Scharf, 2004).

Utilization of light alloys has as main advantages their lightness, this characteristic applied to the transport industry is related directly to the decrease of the pollution by gas emission. Some studies appointed that the reduction of 100 kg in net load lead to a mean reduction of combustible of 0.5 liter/100 km and 2000 kg less of CO_2 production during the mean life of the automobile. The increases in uses of light alloys in parts of automobiles make it much more efficient today if compared with those of the decade of 1970.

Production and application technologies must be cost effective for magnesium alloys to make magnesium alloys an economically viable alternative for the automotive industry. At moment, the energy consumption in light alloys processes production is equivalent to steel or iron alloys.

Close to eighty percent of total energy amount used in processing production of metals were dedicated only for aluminum, iron, copper, zinc and titanium since the mining to the final alloys. A good comparison is when including light alloys recycling process where a necessity of only fifth percent of necessary energy is applied (preservation of mineral).

The environmental benefits provided by lightweight, unlimited, and recyclable Mg alloy have the potential to grow significantly in the future if the global Mg industry is working together to demonstrate its stewardship by eliminating global warming SF_6 or other protective gases as well as Be addition not only for environment and toxicity issues but also for the synergy of cast shop infrastructure with Al industry and by ensuring safety during manufacturing and application, especially without sacrificing process abilities and mechanical properties or increasing the cost of Mg alloy (Kim et al., 2009).

A matrix of potential recycling alloys has been set up to find a potential system for the recycling of heat resistant magnesium alloys. The matrix was processed via permanent mould casting. With the development of new heat resistant magnesium alloys, the automotive industry has introduced several parts to the drive train. The rising number of large magnesium components will result in a higher quantity of automotive post consumer scrap. A matrix of potential recycling alloys based on the Mg-4.97Al-0.3Mn alloy was prepared via permanent mould casting (Fechner, 2009).

The magnesium die casting is a production process which is used for many years for the serial production of different products. The technology used for that has been developed considerably further in the last years. Today the die casting process is a production process with a high grade of automation. Following the state of the art is explained, to describe with it the room for improvement with regard to environment, energy consumption and safety in the process guidance. The described technology allows new procedures so that efficiency and productivity of the die casting process increase. Results are shown by means of some examples, especially in the application of high-grade and thin-walled magnesium die casting (Erhard and Schlotterbeck, 2009).

Therefore the amount of weight of light alloys in transport vehicles which that will need more primary resources could be reduced and all the parts of the environment that would be preserved to the future is huge. At moment only 30% of recycled materials can be used in automotive industry. Notwithstanding of this, the industry announces that more than ninety percent of the vehicle can be recycled today. An environmental problem in automotive industry is the actual life cycle of a vehicle, today is near 3 years.

As mentioned before environmental conservation is one of the principal reasons for the focus of attention on magnesium to provide vehicle weight reduction, CO_2 emission and fuel economy. Weight reduction through Mg applications in the automotive industry is the effective option for decreasing fuel consumption and CO_2 emissions. Improvements in Mg alloying and processing techniques will make it possible for the automotive industry to manufacture lighter, more environmentally friendly, safer and cheaper cars.

Significant research is still desirable on magnesium processing, alloy development, joining, surface treatment, corrosion resistance and mechanical properties improvement to realize future targets to reduce the vehicle mass and the amount of greenhouse gases.

4. Comparative and effective use of new Mg alloys

New light materials are effectively nowadays inserted in world strategies of automotive industry since the environment necessities for pollution and reduction of fuel consumption. Therefore the industry takes part of the risk of development of such alloys but, in fact, some of this has been made at academic level. It's possible to enumerate some aspects of the necessities and characteristics concerning those alloys: low costs, insulation (sound and thermal), impact safety, deformation strength, recyclability and guaranty (to aging as example). All those aspects are linked with the increasing of new automobile models and reflect in production programs more and more complexes.

Different joining techniques were applied to magnesium wrought semi-finished products, in order to promote their introduction on aeronautical structures. Airbus has performed some first tests to join magnesium sheets by friction stir welding. In general the alloy AZ31B (Mg-3.0%Al-0.3%Mn) is quite easily weldable by different processes. Using laser beam welding an AZ61 (Mg-5.9%Al-0.5%Zn-0.2%Mn) filler wire is advantageous for the mechanical properties to weld this alloy.

Another influence in research for new technologies of materials and light alloys relies on the shortening of models life. New materials and alloys are usually more expensive than commercial materials, so there is direct needs of investment in reduce the costs of production and development of such alloys allowing the utilization in large scale in the automobile production processes.

The increase in the potential application of magnesium profiles is strongly dependent on the question of whether established forming processes for aluminum and steel can be changed to magnesium and its alloys. Broad-spectrum applications of magnesium alloys in the automotive industry are casting products.

Despite the fact that the introduction of light alloys and new technology light alloys is a tendency not changeable, the utilization effective today still is more applicable to competition or sportive cars and motorcycles, due to the high costs previously mentioned. However if compared along the existence of automobile the employment of light alloys rise exponentially from the earliest up to the latest commercial model. The influence of 1970's in the development of such technologies is notable comparing with the few kilograms used in the first automobiles.

Recently the weightiness of light alloys, for example, in an automobile is near 90 kg in Europe, 120 Kg in United States and 42 kg in Brazil, but increasing year-to-year. Nearly 90% or more from the weight relies on aluminum alloys, but there is a rapidly increase in the magnesium and a slightly in titanium alloys in the total amount used. All those factors contribute to new researches and development of this class of materials for structural and mechanical applications in automotive industry.

Traditionally the main usage for magnesium and magnesium alloys has been for aluminum alloying, high pressure die casting and steel desulphurization. Over the last 10 years the demand for magnesium and its alloys has grown at an average rate close to 5% per year. The die casting industry which expanded at a rate of over 10% per year was mainly responsible for this steady growth of the whole industry. This remarkable growth was possible because of the stable and relatively inexpensive supply of magnesium from China. This low costs supply has changed at the end of 2007 and early 2008. During that period the base price of magnesium has tripled. In this article present and future opportunity in supply and demand of magnesium and magnesium alloys are examined. Special attention will be given to the growth potential of magnesium alloys for components which will be driven most likely by environmental regulations from governments (Closset, 2008)

As the lightest structural materials, magnesium alloys are well suited for the car industry and also good fuel economy is essential. The selection of a new alloy for a vehicle component should be based on technical requirements and targeted cost. In reality, this selection process is complicated and depends very much on the relative weight given to a specific property, which is part of the combined desired properties and final targeted cost. This task becomes even more complicated if alternative material systems such as aluminum alloys are considered for the same applications.

Several new magnesium alloys have been developed recently for high temperature applications to obtain an optimal combination of die castability, creep resistance, mechanical properties, corrosion performance, and affordability. Most of the new alloys can only partially meet the required performance and cost. The ZE41 alloy (gravity-casting applications) has moderate strength and creep resistance combined with good castability. Although this alloy exhibits poor corrosion resistance, it is still preferred for certain applications.

Although the most commonly used magnesium die-casting alloys are of the AZ and AM series, improved elevated-temperature performance is required (gearbox housing, intake manifolds, oil pans, transfer cases, crankcases, oil pump housing). Insufficient creep strength of alloys can causes poor bearing-housing contact, leading to oil leaks and increased noise and vibration.

The use of magnesium alloy casting in the automobile industry expands at an impressive rate in this decade, which can manage with the energy and environment problems. Alloy

AZ91 (Mg-9Al-0.8Zn-0.2Mn) is the most favored magnesium alloy, being used in approximately 90% of all magnesium cast products (Guangyina et al., 2000).

There are two patented magnesium alloys (Dead Sea Magnesium Ltd, 2012): Mg-Al-Ca-Sr based alloy (MRI 153M) and Mg-Al-Ca-Sr-Sn based alloy (MRI 230D). The MRI 153M is a beryllium-free, creep-resistant alloy capable of long operation at temperatures up to 150°C under high stresses (substantially superior to those of commercial alloys). The MRI 230D is a die-casting alloy developed for use in automotive engine blocks operating at temperatures up to 190°C. The alloy has excellent creep resistance combined with good castability, high strength, and superior corrosion behavior. The results obtained show that MRI 230D and A380 exhibit similar tensile creep behavior at 150–175°C under stress of 70 MPa (Aghion, 2003).

Concerning the whole aeronautic industry, due to the fact that weight reduction is a very important objective for strengthening the competitiveness, several alternatives to obtain weight reduction has to be investigated (welded or bonded airframes; use of metal laminates; structural plastics; fiber reinforced composites).

The non-metallic materials application in selected areas is not conceivable due to restricted properties under low or elevated temperatures, missing electrical conductivity or low damage acceptance. Fiber reinforced plastics are a relatively lavish material only used for primary structure applications with highest demands.

The family of magnesium alloys and especially magnesium wrought materials can be an excellent alternative because of their low density, good mechanical properties, moderate cost and metallic character (in respect of manufacturing, repair, maintenance compared to composites).

In the past decade a lot of research activities and development projects have been carried out working on magnesium cast materials mainly for transport applications. There were only very few activities on magnesium wrought products like sheets, extrusions or forged parts. The alloy spectrum of magnesium wrought alloys is still very restricted.

Aeronautic requirements and applications of wrought products have been evaluated only in a few projects. Increasing the research on magnesium wrought alloys will promote a new class of metallic materials for aeronautical applications to win the competition against plastics and fiber reinforced plastics. Therefore, the variety of offered metallic materials will be enlarged, not only for aircrafts, but also for space, military and satellites applications.

To reach this objective magnesium has to deliver meaningfully higher weight specific mechanical properties compared to aluminum. The aims for aluminum replacement can be divided into two different steps in respect of time scale and risk.

A replacement of medium strength 5XXX aluminum alloys for cockpit and cabin applications and another possible replacement of medium to high strength 2XXX aluminum alloys for secondary structure or non-pressurized fuselage applications.

Forming and joining technologies require development, simulation and validation for the innovative material and technologies commonly used within aeronautic industry. Recently Hombergsmeier presented the requirements of new alloys concerning property temperature systems and structural applications (see Table 1).

Properties	T(K)	Materials	
		Structural applications	Systems applications
Tensile Ultimate Strength	RT	450 Mpa	275 - 350 MPa
Tensile Yield Strength	RT	350 Mpa	200 - 300 MPa
Elongation to fracture	RT	16 - 18 %	12 - 16 %
Yield Strength	423	0.9 YTS	0.9 YTS
Compressive Yield Strength	RT	1.1 YTS to 0.9 YTS	1.1 YTS to 0.9 YTS
Failure under compression	RT	alike: Al 2024 T3	alike: Al 5083
Specific Weight	RT	1.75	1.75
Residual Strength	RT	alike to 2024 T3	n. a.
Fatigue Crack Growth	RT	alike to 2024 T3	n. a.
Fatigue Limit (K_t=1.0, R=0.1)	RT	140 Mpa	160 MPa

Table 1. Requirements for Mg alloys for transport applications (Hombergsmeier, 2009)

The most promising new alloy systems selected due to corrosion behavior, green friendliness and mechanical performance for further investigation as wrought products were Mg-Al-Zn, Mg-Zn-Zr-Re and Mg-Y-Re (Hombergsmeier, 2009, 2011). An undertaking processing for magnesium alloys is the application of severe plastic deformation (SPD) that makes available a useful tool for introducing very significant grain refinement in bulk metallic materials. Even though several SPD techniques are now accessible, processing by equal-channel angular pressing (ECAP) is especially attractive because it is easily scaled-up for use with large samples. Since grain refinement is generally believed to have beneficial effect on properties of Mg alloys, fabrication of bulk ultrafine-grained (grain size less than 1μm) alloys using the new conceptual metal forming process (SPD) should attract considerable attention (Lowe and Zhu, 2003; Valiev, 2004; Figueiredo and Langdon, 2009; Jiang and Ma, 2011).

5. Magnesium alloys for cycle applications

Bicycle frames have gone from wood to steel to advanced alloys to composites in the last century and a half. Bikes themselves have gone from simple curiosities to serious modes of transportation in developing countries and a major form of sports and recreation in developed ones. The improvements in automotive and aerospace industry are now helping to increase a rebirth in the bicycle industry. Similarly, mass production techniques derived for the automotive industry have helped bring the cost down but, many of the original materials and production processes used today remain unchanged from over a hundred years ago (Brower, 2005).

The basic properties of magnesium alloys propose that they would keenly find use in bicycles. Magnesium alloys have low density and a high strength to-weight ratio, are readily extrudable, and some alloys are highly weldable. A lack of information about wrought magnesium alloys, a lack of suppliers to complete the production process, and a lack of information in how to set up a mechanized plant to process wrought magnesium alloys have contributed to only limited market dissemination by magnesium into bicycle applications (Deetz, 2005).

Many doubts limit the recognition of welded magnesium alloy tubes as a substitute bicycle frame material. Most bicycle industry believes that magnesium is not weldable and is very brittle. But some alloys are weldable and it has been recognized that the ductilities of common magnesium alloys (Avadesian, 1999) are equivalent to aluminum alloys, for example, 6061 and 7075 aluminum alloys.

A common misunderstanding is related to wellbeing, with a conviction that contact with oxygen or water will cause explosions. Also that magnesium alloys lack sufficient stiffness to make a bicycle frame. Actually start the acceptance larger diameters to compensate lower stiffness and tensile strength. Concerning the difference between the Young modulus of aluminum and magnesium is much less than the difference between steel and aluminum. With applied basic design principles, it is assumed that the general requirement for product improvements will press on bicycle manufacturers to more seriously consider wrought magnesium as an attractive alternative to the utilized common materials (steel, aluminum, carbon fiber composites).

Magnesium frames may be manufactured by die casting or by welding extruded and welded tubes. Die-casting requires the recovery of die costs that is expensive, but cast magnesium can offer economy with large production volumes. So a manufacturing route could offer a low-cost transportation solution for a developing country.

The wrought magnesium alloys generally offer improved elongation and superior forms. Wrought tubes formed and bent into frame tubes represent a highly flexible and economical approach. Also the weldability rates of magnesium alloys are generally higher than competing aluminum or titanium alloys, or even alloy steels (Welding is the primary method of joining planned for magnesium alloys. His efficiency is very high and many Mg alloys not require post heat treatment).

Among the high damping properties and low density of magnesium it can easily exceed any of the current metals in ride quality leading to an improved fatigue life. Extruded profile designs need to reduce weight and maintain dent resistance in specific key functional areas. Magnesium alloys have a range of potential benefits to offer the bicycle designer, ranging from low density, high specific strength, and dent resistance to extrudability, machinability, and weldability. Finally, making bicycles that win races is the key to providing magnesium a lasting foothold in the cycle industry.

In other areas of the bicycle, rims have been made using Elektron ZM21 and AZ61. Pedals and handlebar stems have also been developed. Wheel hubs, seat posts, brakes, cranks, and ven forks have been produced by companies that specialize in industrialization of high-quality products. Clearly, pioneering an entire class of metal was not well suited to replacement strategy or ad-hoc engineering. Most of the initial efforts have not been successful because rushed planning combined with limited infrastructure ultimately led to problems (Deetz, 1999; Easton et al., 2008, Ashby, 2003).

6. Conclusion

Magnesium has a number of qualities and scantnesses compared to the others lightweight metals. There are a number of developments occurring in the magnesium improvement which promise fine for the future. In the science fundaments there is consciousness that to see the greatest utilization of magnesium both cast and cast and wrought and an adequate scientific foundation is required including new development of deformation behavior in processing to obtain desired microstructures (corresponding good mechanical properties) and near-net shapes (to reduce machining employment); also better phase transformations understanding to allow development of new cast and wrought alloys. The very low density of magnesium together with attractive features like castability is leading to an expanded transport market. Increased use can come from an expanded design base; better understanding of the fundamentals of magnesium behavior and development of cost-affordable cast and wrought alloys.

Author details

W.A. Monteiro
Materials Science and Technology Center - IPEN, São Paulo,
Brazil
School of Engineering - Presbyterian Mackenzie University, São Paulo,
Brazil

S.J. Buso
Uni - Santana, São Paulo,
Brazil

L.V. da Silva
Materials Science and Technology Center - IPEN, São Paulo,
Brazil

Acknowledgement

The authors would like to acknowledge the financial support of IPEN; UPM; CNPq and CAPES (Brazilian government fellowships).

7. References

Aghion, E., Bronfin B, Von Buch, F. (2003). Newly developed magnesium alloys for powertrain applications. *Journal of Metals*, 55, 2003, vol. 11, pp. 30—33.

Avadesian, M. M. and Baker, Hugh (editors), (1999). *Magnesium and Magnesium Alloys*, ASM Specialty Handbook (Materials Park, OH: ASM International), pp. 17.

Ashby M. and Johnson K. (2003). "The Bicycle: Materials and Form", in: *Materials and Design: The Art and Science of Material Selection in Product Design*, Butterworth & Heinemann, pp 108-109.

Beals, R. S., Tissington, C., Zhang, X., Kainer K., Petrillo J., Verbrugge, M. & Pekguleryuz, M. (2007). Magnesium Global Development: Outcomes from the TMS 2007 Annual Meeting, *Journal of Metals*, pp. 39 - 42.

Brower, M. (2005). Advancements in Materials Used in Bicycle Frames, *Term Paper*, (Materials Science and Engineering - EGR250, Instructor: Dr. P. N. Anyalebechi), School of Engineering, Grand Valley State University.

Closset B. (2008). Supply and Demand Developments in the Magnesium Industry, *Proc. 16th Magnesium Automotive and User Seminar,* 18-19 September 2008, Aalen University, Germany.

Davis, B., Wilks, T., Fein, A., Hombergsmeier, E., Entelmann, W. (2009). MagForming - Development of New Magnesium Forming Technologies for the Aeronautics Industry. *Proceedings of IMA Annual World Magnesium Conference,* Fremont, USA, May 31 - June 2, 2009.

Dead Sea Magnesium Ltd, Magnesium alloys for high temperature applications, http://www.dsmag.co.il/?cmd=products.11, accessed in 24 / 03 / 2012.

Deetz, J. (2005). The Use of Wrought Magnesium in Bicycles, *Journal of Metals,* pp 50-53.

Ditze A. (1999). Material cycle for magnesium. *Aluminium.* vol.75, pp. 157— 160

Easton, M., Beer, A., Barnett, M., Davies, C., Dunlop, G., Durandet, Y., Blacket, S., Hilditch, T. & Beggs, P. (2008). Magnesium Alloy Applications in Automotive Structures. *Journal of Metals.* November 2008, Vol. 60, No. 11, 57-62.

Erhard, N. and Schlotterbeck, M. (2009). Die Casting for the Future Economically and Ecologically. *Proceedings of IMA Annual World Magnesium Conference,* Fremont, USA, May 31 - June 2, 2009.

Eliezer, D., Aghion, E. and Froes, F. H. (1998). Magnesium Science, Technology and Applications, in: *Advanced Performance Materials* 5, 201–212.

Fechner, D., Blawert, C., Hort N. & Kainer K. U. (2009). Recycling of magnesium drive train components. *Science in China Series E: Technological Sciences.* vol. 52, 2009, no. 1, pp. 148-154

Figueiredo, R. B. and Langdon, T. G. (2009). Principles of grain refinement in magnesium alloys processed by equal-channel angular pressing. J Mater Sci, vol. 44, 2009, pp. 4758–4762, DOI 10.1007/s10853-009-3725-z

Guangyina, Y., Yangshanb, S. & Wenjianga, D. (2000). Effects of Sb addition on the microstructure and mechanical properties of AZ91 magnesium alloy. *Scripta Materialia,* 43, 2000, pp. 1009–1013

Hombergsmeier, E., (2009). AEROMAG - Magnesium for aerospace applications, in: http://www.materials.manchester.ac.uk/pdf/research/latest/magnesium/elke_hombergsmeier_AEROMAG%20Paper_07.pdf, (22/01/2011).

Hombergsmeier, E., (2011) AEROMAG - Magnesium suitable for aeronautic applications?, *Proceedings of the Sixth European Aeronautics Days – Aerodays,* Madrid, March 30th to April 1st 2011

Kim, S., Lee, J., Jang, D., Yoon, Y., Ha, S., Yoo, H., Park, S., Lee, C., Kim, Y. (2009). Eco-Mg for Magnesium Future, *Proceedings of IMA Annual World Magnesium Conference,* Fremont, USA, May 31 - June 2, 2009.

Kimura, K, Nishii, K & Kwarada, M. (2002). Technology for recycling magnesium alloy housings of notebook computers. *Mater Trans,* 43, 2002, pp. 2516—2522

Kulekci M. K. (2008). Magnesium and its alloys applications in automotive industry. *Int. J. Adv. Manuf. Technol.* Vol.39, 2008, pp. 851–865, DOI 10.1007/s00170-007-1279-2

Knüwer, M., Guillan, A., Besuchet, P., Busch, H.-P., Entelmann, W., Hombergsmeier, E. (2009).Development of Magnesium Alloy Parts for Airbus Aircrafts, Proceedings of IMA Annual World Magnesium Conference, Fremont, USA, May 31 - June 2, 2009

Lowe, T. C. and Zhu, Y. T. (2003). Commercialization of nanostructured metals produced by severe plastic deformation processing. *Adv. Eng. Mater.* vol. 5, 2003, pp. 373–378.

Ostrovsky I., Henn. Y. (2007). Present state and future of magnesium application in Aerospace Industry, *Proceedings of International Conference "New Challenges in Aeronautics",* ASTEC'07, August 19-22, 2007, Moscow

Raynor C. V. (1959). *The Physical Metallurgy of Magnesium and its Alloys,* Pergamon Press, London.

Roberts C. S. (1960). *Magnesium and its Alloys,* Wiley, New York.

Jiang, J. and Ma, A. (2011). Bulk Ultrafine-Grained Magnesium Alloys by SPD Processing: Technique, Microstructures and Properties, chapter 9 in: *Magnesium Alloys - Design, Processing and Properties, Ed. Frank Czerwinski, InTech, ISBN 978-953-307-520-4,* 187 – 219, January 2011.

Scharf C, Blawert C and Ditze A.(2004). Application of remelted post consumer scrap for structural magnesium parts. *Proc. of 6th International Conference Magnesium Alloys and Their Applications*. Weinheim: Wiley-VCH, 2004. 980—987

Valiev, R., (2004). Nanostructuring of metals by severe plastic deformation for advanced properties, *Nature Materials*, www.nature.com/naturematerials, vol. 3, August 2004, pp 511-516.